Cell Polarity in Development and Disease

This Volume is an entry in the

PERSPECTIVES ON TRANSLATIONAL CELL BIOLOGY series

Edited by P. Michael Conn

Perspectives in Translational Cell Biology

Cell Polarity in Development and Disease

Series Editor

P. Michael Conn

Volume Editor

Douglas W. Houston

ACADEMIC PRESS

An imprint of Elsevier

Academic Press is an imprint of Elsevier
125 London Wall, London EC2Y 5AS, United Kingdom
525 B Street, Suite 1800, San Diego, CA 92101-4495, United States
50 Hampshire Street, 5th Floor, Cambridge, MA 02139, United States
The Boulevard, Langford Lane, Kidlington, Oxford OX5 1GB, United Kingdom

Notices
Knowledge and best practice in this field are constantly changing. As new research and
experience broaden our understanding, changes in research methods, professional practices,
or medical treatment may become necessary.

Practitioners and researchers must always rely on their own experience and knowledge in
evaluating and using any information, methods, compounds, or experiments described
herein. In using such information or methods they should be mindful of their own safety and
the safety of others, including parties for whom they have a professional responsibility.

To the fullest extent of the law, neither the Publisher nor the authors, contributors, or editors,
assume any liability for any injury and/or damage to persons or property as a matter of
products liability, negligence or otherwise, or from any use or operation of any methods,
products, instructions, or ideas contained in the material herein.

Library of Congress Cataloging-in-Publication Data
A catalog record for this book is available from the Library of Congress

British Library Cataloguing-in-Publication Data
A catalogue record for this book is available from the British Library

ISBN: 978-0-12-802438-6

For information on all Academic Press publications visit our website at
https://www.elsevier.com/books-and-journals

Working together
to grow libraries in
developing countries

www.elsevier.com • www.bookaid.org

Publisher: Sara Tenney
Acquisition Editor: Sara Tenney
Editorial Project Manager: Fenton Coulthurst
Production Project Manager: Mohanapriyan Rajendran
Designer: Victoria Pearson

Typeset by TNQ Books and Journals

Contents

3. **Cell Polarity and Asymmetric Cell Division by the Wnt Morphogen**

Austin T. Baldwin, Bryan T. Phillips

4. **Cell Polarity in Morphogenesis—Planar Cell Polarity**

Noopur Mandrekar, Baihao Su, Raymond Habas

5. **Polarized Membrane Trafficking in Development and Disease: From Epithelia Polarization to Cancer Cell Invasion**

Erik Linklater, Cayla E. Jewett, Rytis Prekeris

6. **Planar Cell Polarity and the Cell Biology of Nervous System Development and Disease**

J. Robert Manak

7.   Planar Cell Polarity in Ciliated Epithelia

Peter Walentek, Camille Boutin, Laurent Kodjabachian

List of Contributors

Austin T. Baldwin, The University of Iowa, Iowa City, IA, United States

Camille Boutin, Aix-Marseille University, Marseille, France

Tristan Frum, Michigan State University, East Lansing, MI, United States

Raymond Habas, Temple University, Philadelphia, PA, United States; Fox Chase Cancer Center, Philadelphia, PA, United States

Douglas W. Houston, The University of Iowa, Iowa City, IA, United States

Cayla E. Jewett, University of Colorado Anschutz Medical Campus, Aurora, CO, United States

Laurent Kodjabachian, Aix-Marseille University, Marseille, France

Erik Linklater, University of Colorado Anschutz Medical Campus, Aurora, CO, United States

Noopur Mandrekar, Temple University, Philadelphia, PA, United States

Denise Oh, The University of Iowa, Iowa City, IA, United States

Bryan T. Phillips, The University of Iowa, Iowa City, IA, United States

Rytis Prekeris, University of Colorado Anschutz Medical Campus, Aurora, CO, United States

Amy Ralston, Michigan State University, East Lansing, MI, United States

J. Robert Manak, The University of Iowa, Iowa City, IA, United States

Baihao Su, Temple University, Philadelphia, PA, United States

Peter Walentek, University of California, Berkeley, CA, United States

Editor's Foreword

It is an appropriate time to bring together this collection of diverse articles on the role of cell polarity in development and disease. In the ever-growing search for better ways to diagnose, prevent, and treat human diseases, the central role of polarity in the function of many cell types should not be underestimated. In addition to metastatic cancer, which has long been known to arise from changes in cell polarity, work in recent years has revealed the critical role of polarized cilia in many tissues. Also, cell polarity is likely intimately involved in the formation and function of stem cells, germ cells, and their niches, which is of general relevance to the burgeoning field of regenerative medicine. Furthermore, cell polarity is dynamic in nature, and its study benefits from recent advances in imaging and image analysis: so-called "quantitative cell biology."

This book aims to provide scientists interested in going from the cell biological bench to bedside a good background in the basic science surrounding the various aspects of cell polarity, while also introducing the translational relevance. The chapters are organized roughly in developmental sequence, beginning with the egg and stem cells and ending with more differentiated cells. These can be read in any order, but I would recommend the preface "Defining Polarity" by Joseph Frankel for a conceptual introduction to the problem of polarity in biology.

I would like to thank the contributors of the various chapters, Joe Frankel for his insightful preface, Michael Conn for editing this series, and the staff of Elsevier for the work involved in producing this volume.

August 2017   **Douglas W. Houston**
Iowa City, IA, USA

Preface: Defining Polarity

Joseph Frankel,
The University of Iowa,
Iowa City, IA, United States

Cell polarity is a concept that biologists appear to understand without expressing any need for an explicit definition. An exception is the celebrated developmental biologist Lewis Wolpert, who views polarity as "a moderately complex topic" and has sought to define it, most recently in a 2013 "Discussion Meeting" of the Royal Society dealing with "Cellular Polarity: from Mechanism to Disease" (Wolpert, 2013).

Wolpert starts out by asking "is a pencil with a lead point at one end and a little rubber at the other polarized?"

Wolpert's counterintuitive answer is "no." For him, a polar system is not merely one whose ends are different, but one in which it is possible at any point within the system to draw an arrow pointing to one of the poles, a criterion that is met by a magnet, by the body column of a Hydra (Bode, 2011), but not by a pencil ("There are no arrows or asymmetries along the axis of…the pencil"). This latter requirement was expressed more precisely by Francis Crick, who wrote that "a system of polarities…necessarily implies a vector field" (Crick, 1971). Following Crick, I have dubbed this property "vectorial polarity" (Frankel, 2000).

Crick (1971) made an interesting further stipulation: "Whereas one can always derive a vector field from a scalar one, the converse is not always true." Stated differently, while the existence of a gradient in a biological system implies the presence of vectorial polarity in that system, the existence of vectorial polarity need not require the presence of an underlying gradient. As I have argued elsewhere, this latter condition (polarity in the absence of a relevant gradient) applies to the propagation of the ordered cytoskeletal framework of ciliary rows in ciliates such as *Paramecium* and *Tetrahymena* (Frankel, 2000). It could apply elsewhere.

If one surveys the ever-growing literature in this field, one finds that two types of polarity are frequently invoked: planar and basal–apical. Taken literally, these terms apply to the geometrical settings in which polarity is manifested, rather than to the nature of the polarity itself. To find out how polarity is actually defined in these two contexts, we need to dig more deeply into the situations in which these terms ("basal–apical" and "planar") are utilized by biologists. I will consider planar polarity first and basal–apical polarity second.

The paradigm for planar polarity is the uniform polarity of the trichomes on the wing of *Drosophila melanogaster.* This involves both "polarized cell—cell interactions that align cells with their immediate neighbors and long-range patterning events that orient this polarization with the axes of the tissue" (Goodrich and Strutt, 2011, p. 1877). The short-range cell—cell interactions have been beautifully worked out: they involve primarily a "core complex" of six proteins. The principal players are the complementary transmembrane proteins Frizzled and van gogh (a.k.a. Strabismus), backed up by the symmetrical atypical cadherin Flamingo (a.k.a. Starry Night) and three associated cytosolic proteins, Diego, Dishevelled, and Prickle. Goodrich and Strutt (2011) describe the results of experiments in several labs that have established that the direction of polarity in the pupal wing disc is determined by local interactions between Frizzled and its supporters (Dishevelled and Diego) at the distal edges of epidermal cells, interacting with van gogh and its main supporter (Prickle) at the proximal edges of adjacent cells.

Although the nature of the short-range polarization across cell boundaries in the pupal wing disc of *D. melanogaster* has (apart from a few messy details) been substantially solved, the same cannot be said for the long-range coordination of tissue polarity. According to Goodrich and Strutt, "One of the least understood aspects of planar polarity is how cells align along a specific tissue axis" (Goodrich and Strutt, 2011, p. 1887), a situation that apparently still exists. There have been several proposals, with varying support, whose evaluation lies outside the scope of this brief preface. Thus, while the existence of a system that supports the vectorial polarity of adjacent cells on the *Drosophila* wing disc is beyond question, whatever it is that causes *all* the trichomes on the wing surface to point in the *same* direction is still a work in progress. A source of uniform vectorial polarity should be found somewhere in this system. Nonetheless, the remote possibility that, as in ciliates, this uniform vectorial polarity is achieved without governance by a long-range gradient has not been ruled out. However, unlike the situation in ciliates, whatever is at work to orient future trichomes in the *Drosophila* pupal wing disc must work at least in part between, rather than within, cells. For additional background information as well as reviews on the roles of planar cell polarity proteins in vertebrate morphogenesis and neural development, please see the chapters herein by Mandrekar et al. (2018) and Manak (2018), respectively.

The concept of "planar cell polarity" rapidly spread beyond its initial realm of *Drosophila* wing trichomes (and the compound eye ommatidia). One of the key areas of its application, not available in *Drosophila,* is the uniform planar orientation of the clusters of cilia on the surface of ciliated epithelia. This is the topic of the chapter by Walentek et al. (2018) in Chapter 7. Homologs of the key players in *Drosophila* planar polarity have major roles in the orientation of beating cilia, which I will leave the reader to discover in the Walentek article and elsewhere. I will add little here, except for a terminological guide to the names of orthologous genes, plus one old story that I find fascinating.

The Terminological Guide

Drosophila	Vertebrate
fz (frizzled)	Fzd1, Fzd2, Fzd3, Fzd6 (Frizzled 1,2,3,6)
stbm (strabismus)/vang (van gogh)	Vangl1,Vangl2 (VANGL planar cell polarity protein 1,2)
dsh (dishevelled)	Dvl1, Dvl2, Dvl3 (Dishevelled 1,2,3)
pk (prickle)	Prickle 1,2,3
fmi (flamingo)/stan (starry night)	Celsr 1,2,3 (Cadherin EGF LAG seven-pass G-type receptor 1,2,3)

Modified from Goodrich, L.V., Strutt, D., 2011. Principles of planar polarity in animal development. Development 138, 1877—1892.

Not covered in the Walentek article is an experiment conducted by Boisvieux-Ulrich and Sandoz over 25 years ago (1991), on the quail oviduct, which is lined by multiciliated cells devoted to conveying the ovum from the infundibulum of the fallopian tube to the uterus. All of its cilia beat in the same direction toward the uterus. The experiment consisted of inverting a segment of immature oviduct long before the onset of ciliogenesis, which occurs in response to estrogen and after extensive proliferation. As a consequence, the direction of beating of the cilia of the rotated segment was reversed, and the orientation of the "basal feet" and rootlet structures near the bases of the cilia were correspondingly inverted. The authors concluded that "the polarized organization of the ciliary pattern depends on a factor present in the primary cells of the immature oviduct, transmitted through a long series of divisions, and eventually expressed in the final stage of division" (Boisvieux-Ulrich and Sandoz, 1991, p. 10). A ciliatologist like myself would wonder whether what is being transmitted has a structural basis that in any way resembles the cytoskeletal framework of a ciliary row, which in *Paramecium* can also be propagated following experimental inversions (Beisson and Sonneborn, 1965). Thus, whereas ciliary polarity in the frog epidermis seems to be coupled to mechanical strain arising during morphogenesis (Chien et al., 2015), it is unclear whether this accounts for polarity in tubular ciliated epithelia. Unfortunately, the principal investigator of this project (Daniel Sandoz) passed away shortly after his 1991 article was published, and apparently no one else has returned to cell polarity in the quail oviduct to solve its mysteries.

Returning now to our central question, we have seen that the phenomenon called "planar cell polarity" clearly passes the Wolpert/Crick test for vectorial polarity. Planar cell polarity is conceptually more like a magnet than a pencil. Nonetheless, in some of the best-known planar cell polarity systems, the question of whether (and how) the vectorial polarity is driven by a scalar field (such as a Wnt gradient) is still unresolved.

Basal–apical polarity, in my view, is something entirely different. It is excellently characterized in the article by Linklater et al. in Chapter 5. The key observation is that in a basal–apically polarized system, "Individual epithelial cells are polarized to maintain the identity and distinct compositions of the apical and basolateral sides of the cell" (Linklater et al., 2018). In its paradigmatic system, the transport epithelium of the vertebrate kidney, an apical compartment faces the lumen and promotes entry of sodium ions and solutes by way of specific ion channels and transporters, while a basal (or basolateral) compartment faces the animal's (topological) interior where sodium ions are actively pumped out of the epithelial cell by Na^+/K^+ ATPases (Nelson et al., 2000). In addition, "The apical and basolateral domains are separated by a group of scaffolding proteins that form a structure called the tight junction…[that]…acts as a diffusion barrier to prevent mixing of apical and basolateral membrane components, and also functions as an intercellular seal" (Linklater et al., 2018, citing Zihni et al., 2016).

Can such a system be considered polarized? If so, where, if anywhere, can one draw an arrow of vectorial polarity? It seems to me, nowhere, except perhaps at the boundary between the apical and basolateral membrane domains. Polarization exists primarily between rather than within these domains. So I would call this "domain polarity," as distinct from the vectorial polarity that is a hallmark of planar cell polarity. I have earlier, also following Wolpert (1971), made this distinction for two superimposed systems of polarity in ciliates (Frankel, 2000) and now find it equally applicable to the animal systems described in this book.

Domain polarity also exists in places quite different from its paradigmatic location in secretory epithelium. Frum and Ralston (2018), in Chapter 2, connect compartmental organization to developmental potency in mouse embryos. Up to and just after the third division of the embryo, the eight blastomeres are separate, with spaces between them, and most or all of them have retained totipotency; i.e., the capacity to generate all tissues of the organism as well as extraembryonic tissues such as the placenta. But then, before the fourth division, the embryo undergoes compaction, and the spaces between the cells disappear. At this point, molecular indicators of "polarization" appear at the outer surface of some of the cells. "By the time the eight-cell embryo has completed the process of compaction, each blastomere contains a molecularly distinct apical domain on the embryo's surface, and a molecularly distinct basolateral domain at the sites of cell–cell contact inside the embryo" (Frum and Ralston, 2018). The blastomeres now exhibit classical domain polarity in an unusual context.

Then, at the fourth division, something strange happens. Depending on the orientation of the cleavage plane, some cells sink to the interior (becoming the beginning of the inner cell mass), while others maintain their contact with the cell surface. The cells that preserve contact with the cell surface retain their previous domain polarity and later are forced to choose between contributing further to the inner cell mass or differentiating into cells of the trophectoderm, where they are restricted to extraembryonic fates. Those cells that become internalized become nonpolar (because they now have only a single domain), and pluripotent, retaining the capacity to produce all of the cell types of the organism while losing the capacity to form extraembryonic tissues.

This, in my view, is truly a remarkable situation, in that embryonic cells that start out with two polar domains can generate daughter cells that become nonpolar simply by retaining just one of the two polar domains. This type of loss of polarity could not occur in a vectorial system, in which the polarity is expressed at every point within the system, as in a magnet. Another example of this type of asymmetric cell division is described by Baldwin and Phillips (2018) in Chapter 3, although in this case the polarity is determined by secreted Wnt proteins.

Terminology in biology has a curious sticking power, as we can see in the retention of alternative gene names. Naïve outsiders can read one paper about *strabismus* and another about *van gogh* and think that these are different genes. That is an example in which two terms refer to one thing. In the case of basal—apical and planar polarity, one term (polarity) is used to refer to two things: vectorial polarity in a "planar" system and domain polarity in a "basal—apical" system. As Wolpert wrote, cell polarity is indeed a "moderately complex topic."

REFERENCES

Baldwin, A.T., Phillips, B.T., 2018. Cell polarity and asymmetric cell division by the Wnt morphogen. In: Houston, D. (Ed.), Cell Polarity in Development and Disease. Elsevier, Amsterdam, pp. 61—91.

Beisson, J., Sonneborn, T.M., 1965. Cytoplasmic inheritance of the organization of the cell cortex in *Paramecium aurelia*. Proc. Natl. Acad. Sci. U.S.A. 53, 275—282.

Bode, H., 2011. Axis formation in Hydra. Annu. Rev. Genet. 45, 105—117.

Boisvieux-Ulrich, E., Sandoz, D., 1991. Determination of ciliary polarity precedes differentiation in the epithelial cells of quail oviduct. Biol. Cell 72, 3—14.

Chien, Y.H., Keller, R., Kintner, G., Shook, D.R., 2015. Mechanical strain determines the axis of planar polarity in ciliated epithelia. Curr. Biol. 25, 2774—2784.

Crick, F.H.C., 1971. The scale of pattern formation. In: Davis, D.D., Balls, M. (Eds.), Control Mechanisms of Growth and Differentiation. (25th Symposium of the Society for Growth and Differentiation). Cambridge University Press, Cambridge, p. 429.

Frankel, J., 2000. Cell polarity in ciliates. In: Drubin, D. (Ed.), Cell Polarity. Oxford University Press, New York, pp. 78—105.

Frum, T., Ralston, A., 2018. Pluripotency—what does cell polarity have to do with it? In: Houston, D. (Ed.), Cell Polarity in Development and Disease. Elsevier, Amsterdam, pp. 31—51.

Goodrich, L.V., Strutt, D., 2011. Principles of planar polarity in animal development. Development 138, 1877—1892.

Linklater, E., Jewett, C.E., Prekeris, R., 2018. Polarized membrane trafficking in development and disease: from epithelia polarization to cancer cell invasion. In: Houston, D. (Ed.), Cell Polarity in Development and Disease. Elsevier, Amsterdam, pp. 121−139.

Manak, J.R., 2018. Planar cell polarity and the cell biology of nervous system development and disease. In: Houston, D. (Ed.), Cell Polarity in Development and Disease. Elsevier, Amsterdam, pp. 147−169.

Mandrekar, N., Su, B., Habas, R., 2018. Cell polarity in morphogenesis—planar cell polarity. In: Houston, D. (Ed.), Cell Polarity in Development and Disease. Elsevier, Amsterdam, pp. 103−114.

Nelson, W.J., Yeaman, C., Grindstaff, K.K., 2000. Spatial cues for cellular asymmetry in polarized systems. In: D.Drubin (Ed.), Cell Polarity. Oxford University Press, New York, pp. 106−140.

Walentek, P., Boutin, C., Kodjabachian, L., 2018. Planar cell polarity in ciliated epithelia. In: Houston, D. (Ed.), Cell Polarity in Development and Disease. Elsevier, Amsterdam, pp. 177−202.

Wolpert, L., 1971. Positional information and pattern formation. Curr. Top. Dev. Biol. 6, 183−224.

Wolpert, L., 2013. Cell polarity. Philos. Trans. R. Soc. B 368. https://doi.org/10.1098/rstb.2013.0419.

Zihni, C., Mills, C., Matter, K., Balda, M.S., 2016. Tight junctions: from simple barrier to multifunctional molecular gates. Nat. Rev. Mol. Cell Biol. 17, 564−580.

Chapter 1

Cell Polarity in Oocyte Development

Denise Oh, Douglas W. Houston
The University of Iowa, Iowa City, IA, United States

INTRODUCTION

Polarity is a central feature of many cell types. Early cell biologists recognized that cells can have polarized arrangements of components, conferring different functional or metabolic properties in different regions. This polarity was evident, even with primitive microscopes, in the lumenal and ablumenal regions of tubule epithelial cells, various gland cells, neurons, and notably, in the germ cells (Wilson, 1928). The egg became particularly important in this regard because of the large size and accessibility of eggs in many species as well as the fortuitous presence of various pigments, which allowed the differential fates of egg regions to be reliably traced. In many invertebrates, the presence of a variety of "crescent" cytoplasms in different regions was closely correlated with the direction of differentiation in those regions and gave rise to the idea of "determinants" housed in the cytoplasm. This localization of determinants is especially evident in the five cytoplasmic regions of the ascidian egg, famously characterized by Conklin (1905). Earlier cytoplasmic ablation studies in ctenophores and mollusks showed that different cytoplasmic regions were indeed functionally important, leading to loss of specific structures when removed. In vertebrates, the amphibian egg was long-noted for its pigmented animal region and pale vegetal region, each with a propensity to contribute to neural and sensory structures or gut, respectively (Wilson, 1928).

One of the enduring examples of polarity in the eggs is the localization of the so-called germ plasm (pole plasm in insects, P granules in nematodes) to one pole of the egg (Houston and King, 2000). Dense cytological "granules" were found continuously in the germ line of copepods and insects, in early studies, and later in frog eggs (Hegner, 1914; Huettner, 1923; Bounoure, 1931, 1934). These granules accumulated in the posterior pole of the insect egg and vegetal cells in the frog egg and were incorporated into the forming pole cells (primordial germ cells, PGCs), and physical ablation or ultraviolet irradiation

Cell Polarity in Development and Disease. https://doi.org/10.1016/B978-0-12-802438-6.00001-2

of these regions generally led to sterility (Geigy, 1931; Bounoure, 1937). Importantly, transplantation of the pole plasm could direct the ectopic formation of functional pole cells at the anterior pole (Illmensee and Mahowald, 1974), providing the first best example of polarized developmental determinants in the egg. Only recently, however, has it been shown that ectopic germ plasm may induce ectopic PGCs in the animal region of the frog embryo (Tada et al., 2012).

In addition to the germ plasm, other important developmental activities were found regionalized to the vegetal hemisphere, including cortical determinants important for the formation of the dorsoventral axis (Ancel and Vintembenger, 1948) and others for the ability of vegetal cells to induce mesoderm in the animal and equatorial regions (Nieuwkoop, 1969; Sudarwati and Nieuwkoop, 1971). Although these early examples in vertebrates came from experiments in amphibian embryos, the polarization of determinants is likely important for germ cell and germ layer formation across vertebrate taxa (reviewed in Houston, 2017). Notable exceptions include the lack of a vegetal germ plasm in urodele amphibians (salamanders/newts (Sutasurya and Nieuwkoop, 1974; Nieuwkoop and Sutasurya, 1976)) and the overall lack of egg polarity with respect to developmental determinants in mammals (Gardner, 1998). Thus, the polarization of the egg as a mechanism to establish differential cell fate in early blastomeres is an ancestral feature of animal development that was likely specifically lost during mammalian evolution.

In many organisms, definitive egg polarity is determined during the early stages of oogenesis and is often related to the formation of a visible "mitochondrial cloud" or Balbiani body (Bb) (Fig. 1.1). This structure was first seen in the oocytes of spiders and centipedes and is a general feature of oogenesis in most organisms (Wilson, 1928; Kloc et al., 2004b). In frogs and fish, the Bb is

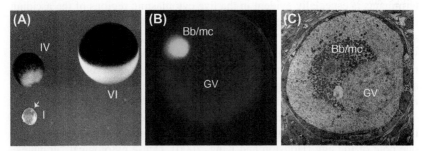

FIGURE 1.1   **Polarity in oocytes.** (A) *Xenopus* oocytes at different stages of animal—vegetal polarization; stages are indicated in Roman numerals, the mitochondrial cloud (mc)/Balbiani body (Bb) is indicated by an arrow. (B) Fluorescent image of a stage I oocyte labeled for mitochondria, showing accumulation in the mc/Bb. *GV*, germinal vesicle (nucleus). (C) Electron micrograph of a human oocyte showing the ultrastructure of the Bb. *From Hertig, A.T., 1968. The primary human oocyte: some observations on the fine structure of Balbiani's vitelline body and the origin of the annulate lamellae. Am. J. Anat. 122, 107–137. https://doi.org/10.1002/aja.1001220107.*

one of the earliest manifestations of animal—vegetal polarity and indeed likely has a role in its establishment. The Bb undergoes breakdown during oogenesis and its constituents are transported to the future vegetal pole (Dumont, 1972; Selman et al., 1993). One of the consequences of this establishment of polarity is the localization of specific mRNAs to the vegetal pole. Many of these associate with the early Bb as part of the germ plasm, and others use a later localization pathway established by the Bb transport machinery (Forristall et al., 1995; Kloc and Etkin, 1995). In some notable cases, the function of individual RNAs has been characterized, and generally these functions correlate with known roles of the vegetal pole (i.e., germ layer formation and patterning and PGC specification; reviewed in Houston, 2013).

The egg is also polarized in a more local sense with regard to the potential for sperm entry. The site of fertilization is often restricted to the animal pole, defined as the site of polar body extrusion during meiosis. In insects and teleost fish, fertilization occurs via the animally located micropyle or tunnels through the follicle cells, which aids in sperm attraction and facilitates entry to the egg (Amanze and Iyengar, 1990; Murata, 2003). Thus, one consequence of cell polarity in the egg is competence for fertilization and the initiation of development. This requirement is likely also true in mammals, suggesting this local polarization can occur independently of overall animal—vegetal polarity. Also, in organisms such as fish and frog, inheritance of polarized egg membrane by outer blastomeres establishes a polarized superficial epithelial lineage. Similar mechanisms of cortical polarization occur in the mammalian morula, but in contrast to the aquatic animals, this occurs de novo and leads the outer cells to form the extra embryonic trophectoderm. Thus, the establishment of polarity either within the egg cell or around its circumference can have important consequences for the developmental fate of early blastomeres. The localization and roles of localized mRNAs in vertebrate eggs and oocytes have been well reviewed recently (reviewed in Houston, 2013; Escobar-Aguirre et al., 2017); here we will give an overview of the molecular mechanisms for the origination of oocyte polarity in model organisms. Many of the themes in this chapter will be reflected in later chapters in this volume but we focus on the oocyte and egg in which the establishment of polarity sets the stage for many aspects of polarity later in development.

OOGENESIS AND THE ESTABLISHMENT OF ANIMAL—VEGETAL POLARITY

The animal—vegetal cytoarchitecture of the oocyte is gradually established during oogenesis. During this time, the oocyte remains arrested in diplotene of Meiosis I until ovulation. In *Xenopus*, where this process has been well-described, the oocyte accumulates yolk and greatly increases in volume and undergoes cytoskeletal and other changes in preparation for eventual ovulation and fertilization (Rasar and Hammes, 2006). In mammals, oocytes are arrested in diplotene but do not undergo progressive growth until after puberty, and then only a small subset begins growing and can be ovulated if this occurs during the appropriate phase of the ovarian hormone cycle (Pepling, 2006).

In *Xenopus*, the animal−vegetal axis can be identified in early oocytes by the translocation of the mitochondrial cloud/Bb to the presumptive vegetal pole. Oocytes begin accumulating yolk and become opaque during stage II and then acquire pigment uniformly at stage III. Pigment distribution becomes polarized during stage IV, generating the stereotypical external appearance of the amphibian oocyte (Dumont, 1972; Rasar and Hammes, 2006). Additionally, the germinal vesicle (GV) (nucleus) becomes positioned in the animal hemisphere during this period. Cytoskeletal organization becomes polarized along the animal−vegetal axis as well (Wylie et al., 1985; Gard, 1999; Sardet et al., 2002; Houston, 2013). Data have suggested that the actin cytoskeleton is not heavily polarized in the growing oocyte, although actin is abundant in the GV (Gard, 1999). The role of this nuclear actin is unclear, but recent experiments suggest possible roles in chromatin remodeling related to pluripotency (Miyamoto et al., 2011, 2013). Keratin-intermediate filaments have been implicated in maintaining some aspects of animal−vegetal polarity. These proteins become differentially organized along the animal−vegetal axis around stage IV. In the vegetal pole, keratin forms a thicker network in an intricate geodesic pattern (Klymkowsky et al., 1987), whereas the network in the animal pole is thinner and more fine-grained. The functional consequences of this difference in organization are not known, but it has been suggested that organized keratin maintains or anchors organelles and localized mRNAs to the vegetal cortex (Alarcón and Elinson, 2001).

As with many cell types, the main cytoskeletal system involved in oocyte polarity is the microtubule cytoskeleton. Microtubules are enriched in the animal pole, where they are acetylated and associate with the GV. Microtubule disruption can disposition the GV, suggesting a role for maintaining the GV in the animal hemisphere. Vegetally, microtubules are found associated with clusters of endoplasmic reticulum in the cortex (Gard, 1991). Importantly, gamma tubulin, a critical component of microtubule-organizing centers, is polarized. Gamma tubulin puncta are present throughout cortex in the vegetal hemisphere (Gard, 1994). The role of these foci is unclear, but they may be involved in the bulk of microtubule nucleation, since oocytes lack centrosomes and thus global regulation of microtubule organization. Additionally, and consistent with the cortical location of gamma tubulin foci, microtubule organization in the *Xenopus* oocyte has the majority of minus ends oriented toward the cortex, which is the opposite to most somatic cell types (Pfeiffer and Gard, 1999).

RNA localization is one important consequence of animal−vegetal polarization during oogenesis. Germ plasm-associated RNAs accumulate within the Bb during stage I, but the majority of localized RNAs do not begin to be localized until stage III, when animal—vegetal polarity begins to be elaborated (Kloc et al., 2002; King et al., 2005). This was first observed for the first localized RNA described, *vg1/gdf1*, and has since been found for many other RNAs and has been described as a "late pathway" of localization. The localization mechanisms appear to begin during stage III, becoming most active in stage IV, and then persisting but with lessened activity in full-grown oocytes (Yisraeli et al., 1990; Alarcón and Elinson, 2001). RNA localization to the Bb is thought to occur independently of cytoskeletal regulation, but the late pathway requires intact microtubules (Zhou and King, 1996; Kloc et al., 1996; Chang et al., 2004). Interestingly, kinesin plus end motor activity is required for vegetal localization—a somewhat paradoxical finding considering the overall microtubule organization of the oocyte (Betley et al., 2004; Yoon and Mowry, 2004; Messitt et al., 2008). However, several groups have found that a substantial subpopulation of microtubules ($\sim 20\%$) have plus ends oriented vegetally during these localization stages (Pfeiffer and Gard, 1999; Messitt et al., 2008), providing a likely explanation for the kinesin-dependent localization of mRNAs opposite to that of general microtubule polarity in the oocyte. Interestingly, in zebrafish, many homologs of the late pathway RNAs are localized animally in the oocyte (Howley and Ho, 2000), suggesting that analogous subpopulations of differentially oriented microtubules might be a general feature of oocytes in which RNA localization occurs.

Thus, in *Xenopus* and likely many other animals, animal—vegetal polarity is gradually elaborated during oogenesis and involves irreversible changes to the cytoskeleton and the distribution of localized mRNAs and proteins. Animal—vegetal polarity is not externally obvious in mammalian oocytes, but a polarization does occur during oocyte maturation when the GV migrates toward a presumptive "animal pole" and undergoes asymmetric cell division resulting in the budding of the first polar body (Maro et al., 1986; Van Blerkom and Bell, 1986; Verlhac et al., 2000). This event thus defines the animal pole according to the commonly accepted definition (i.e., the site of polar body emission in the egg) and is thought to induce regional apical-like polarity in the egg, including the formation of a microvilli-free zone and the accumulation of polarity proteins Pard6 and Pard3 (Maro et al., 1986; Van Blerkom and Bell, 1986; Vinot et al., 2005). Fertilization tends to occur in this animal region (although not in the microvilli-free zone; see below (Hiiragi and Solter, 2004)), suggesting there is some polarity information imparted to the egg; evidence would suggest that this polarity does not contribute to specifying cell fates in the embryo. Cytoplasmic ablation experiments show that eggs lacking animal or vegetal cytoplasm can develop normally (Ciemerych et al., 2000), and the only localized mRNAs identified in the mammalian egg and early embryo are those also associated with the polar body and Meiosis II spindle (VerMilyea et al., 2011). Additionally, cell fates in the mouse blastocyst are primarily established by later polarization of outer blastocyst cells and regulative interactions (see Chapter 2 (Rossant and Tam, 2009; Zernicka-Goetz et al., 2009)).

In *Xenopus* and zebrafish, animal−vegetal polarity is established through the formation and subsequent translocation of the Bb to the cortex, thereby defining the vegetal pole of the oocyte (see below). Mouse oocytes transiently form a Bb (Pepling et al., 2007), and although unlikely, it is not known to what extent the side opposite to the Bb forms the eventual animal pole. The formation of the Bb in mammals may therefore represent a transient asymmetry in the oocyte that is not converted into relevant functional polarity for fertilization or for later development.

ORIGIN OF ANIMAL−VEGETAL POLARITY DURING OOGENESIS

Many key steps in oocyte and germ cell development are conserved in vertebrate and invertebrate animals, including PGC migration to the gonad and differentiation of oogonia through a series of stereotyped oogonial cell divisions. These cleavages result in incomplete cytokinesis, forming a cyst (or "nest") of 16 oocytes that remain interconnected by intercellular bridges. This process has been well-studied in this regard in Drosophila; however, in this case only one "cystocyte" develops into the oocyte, and the remaining cells differentiate into nurse cells (Pepling et al., 1999). By contrast in vertebrates, all cyst cells differentiate into oocytes. Thus, it is thought that oocyte fate determinants are asymmetrically distributed in Drosophila, whereas the process is less restrictive in vertebrates, a mechanistic distinction that is not well understood. In Drosophila, the fate of the oocyte and the polarity of the cyst are regulated by the fusome, an agglomeration of organelles and structural proteins (e.g., spectraplakins) that anchors and penetrates the intercellular bridges (Grieder et al., 2000).

Definitive animal−vegetal polarity in the oocyte requires the formation of the Bb, or mitochondrial cloud, following entry into meiosis. Although the importance of the Bb has been long recognized, it has been mysterious to what extent earlier asymmetries direct the polarized positioning of the Bb and thus eventual animal−vegetal polarity. Oogonial cells in the cyst are often described as "pear-shaped" and are polarized with the centrosomes and clustered mitochondria in the narrow end and the nucleus in the wider end, an arrangement thought to be carried over from the PGCs (al-Mukhtar and Webb, 1971; Coggins, 1973). During oogonial cyst cell divisions, which are typically synchronous, the narrow ends cluster together in the center of the cyst, forming the so-called "rosette" arrangement. Recent evidence in zebrafish suggest that microtubules persist through the final division canal and connect the centrosomes in adjacent cells, along with the mitochondrial clusters (Elkouby et al., 2016) (Fig. 1.2), thus suggesting a cellular mechanism for aligning cyst divisions with the presumptive animal−vegetal axis.

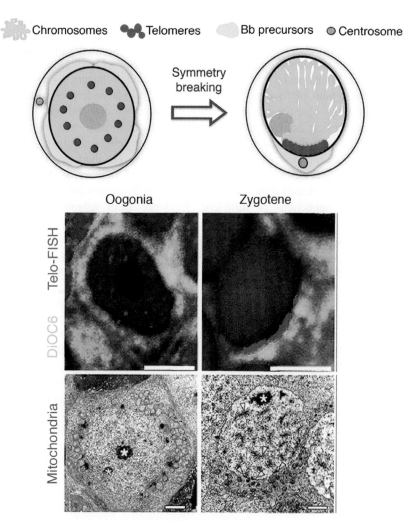

FIGURE 1.2   Origin of the Balbiani body (Bb) in zebrafish. Top, model for symmetry breaking in early meiosis. Mitochondria (DiOC6, green) cluster with telomeres (Telo-FISH, red) of the chromosomal bouquet and the maternal centrosome (orange). Middle, fluorescent images of oogonia and zygotene-stage oocytes showing this clustering. Bottom, electron micrographs showing positions of mitochondria (purple) at the oogonia and zygotene stages. *, nucleoli; *arrowhead*, synaptonemal complexes. *Images were modified and adapted from Elkouby, Y.M., Jamieson-Lucy, A., Mullins, M.C., 2016. Oocyte polarization is coupled to the chromosomal Bouquet, a conserved polarized nuclear configuration in meiosis. PLoS Biol. 14 (1), e1002335. https://doi.org/10.1371/journal.pbio.1002335 under the terms of the Creative Commons Attribution License CC BY 4.0.*

Xenopus

The origin of the Bb/mitochondrial cloud has been described morphologically in *Xenopus* (al-Mukhtar and Webb, 1971; Coggins, 1973; Heasman et al., 1984). Soon after oocytes enter meiosis, mitochondrial clusters disperse uniformly throughout the oocyte. Additionally, the oocyte undergoes folliculogenesis, during which the cyst structure breaks down and the intracellular bridges connecting oocytes are eliminated. Somatic follicle cells invest the oocyte and likely facilitate the breakdown of intracellular bridges, as was recently shown for Drosophila spermatogonial cysts (Lenhart and DiNardo, 2015). Overall, oocytes tend to lose morphological polarity at this stage, although it is not clear if signals from the follicle cells are involved or if loss of intercellular connections is sufficient. Evidence from zebrafish suggests however that signaling between the oocyte and follicle cells is critical for establishing overall animal–vegetal polarity (see below). Centrosome inactivation likely occurs at the onset of folliculogenesis (Gard et al., 1995), representing another manifestation of loss of polarity, although this phenomenon is not well understood.

During later stage I in *Xenopus*, mitochondria accumulate into small perinuclear clusters, "premitochondrial clouds," and begin to accumulate localized mRNAs (Kloc et al., 1996). Eventually, germ plasm is assembled in one of the mitochondrial aggregates, and this structure begins to develop into the definitive mitochondrial cloud (Heasman et al., 1984; Kloc et al., 2004a). The Bb continues to grow along with the oocyte, and by later stage I (~ 250 μm in diameter), the cloud begins to become fragmented and translocated to the presumptive vegetal pole, an irreversible event that essentially determines the vegetal pole of the oocyte. However, live observation suggested that the Bb is immobile within the oocyte (Heasman et al., 1984), indicating that the animal–vegetal axis is likely structurally determined by the process of Bb formation.

Zebrafish

Recent work in zebrafish has extended these studies, showing that nuclear polarity early in meiosis likely determines the position of the Bb and hence animal–vegetal polarity. Confocal analysis of localized proteins and RNAs during the onset of meiosis in zebrafish showed that Bb formation (marked by precursor proteins and RNAs) coaligned with nuclear polarity established by the "zygotene chromosomal bouquet" a cluster of telomeres and synapsed chromosomes in meiosis (Elkouby et al., 2016). This structure has been noted as a near universal feature of meiosis and is thought to facilitate homologous chromosome pairing and recombination (Zickler and Kleckner, 1998; Harper et al., 2004). The bouquet forms by attachment of telomeres to the nuclear envelope and subsequent clustering toward one side of the nucleus. This process is microtubule-dependent and likely involves linkage to transnuclear envelope proteins (see below) and to centrosomes (Elkouby et al., 2016). Other than acting as a symmetry-breaking mechanism, possibly by increasing local concentration of critical components, such as the Buckyball protein (see below (Heim et al., 2014; Riemer et al., 2015)), the coupling of Bb formation to a meiosis-specific event may ensure that the Bb only forms in meiotic cells.

Slightly later in meiosis, the nucleus develops a concavity, or "cleft" (Elkouby et al., 2016) at the bouquet site and partially envelops the centrosome and Bb material. The significance of the cleft is unclear but resembles the classically described "nuclear bay" that houses the centrosome in multiple cell types (Wilson, 1928). Recent observations have suggested the bay may be site for proteasomal degradation of proteins at the centrosome, but it is not known whether proteasome function is needed for oocyte polarity or for Bb assembly. The nuclear cleft is transient and both the nucleus and Bb become spherical as the oocyte undergoes its major growth phase. Thus, animal–vegetal polarity arises from a nuclear symmetry-breaking event in meiosis and a coupling of those events to the recruitment of Bb components in the cytoplasm.

Mouse

Although animal–vegetal polarity (and other forms of oocyte polarity) is established in many organisms, mammalian eggs are notable for lacking substantial polarity. Some local polarity is set up surrounding the Meiosis II spindle and is related to regionalizing sperm entry, but embryological experiments have generally ruled out the effect of oocyte/egg polarity on subsequent mammalian development. During the evolutionary history of mammals, egg development underwent dramatic alterations, likely transitioning from a large, yolk-laden, discoidally cleaving egg to a smaller yolk-free egg exhibiting holoblastic cleavage. Additionally, the first cell divisions became devoted to generating the tissue layers needed for implantation into the uterus. Thus, it is perhaps not surprising that animal–vegetal developmental information no longer exists in mammals. Despite this, the early events in mammalian oogenesis are largely similar to other organisms.

As in *Xenopus* and zebrafish, mouse oocytes initially develop within germ line cysts during embryonic development (~E10.5–E13.5 (Pepling and Spradling, 2001)). However, these cysts are generally lost prior to birth, accompanied by a large increase in germ cell apoptosis (Pepling and Spradling, 2001). As in other vertebrates, the cyst divisions generate only oocytes, not nurse cells per se. However, recent in vivo lineage tracing data suggest that mouse cysts partially fragment and recombine with each other, producing chimeric cysts derived from different PGCs (Lei and Spradling, 2013a). The number of cysts approximates the eventual number of follicles and subsequent studies showed that cyst cells can contribute organelles and other material to the presumptive oocyte, acting as "nurse-like" cells (Lei and Spradling, 2016) analogous to Drosophila. Much of this organelle material contributes to the Bb in the forming oocyte, suggesting that healthy organelles, associated localized RNAs, and nuage material are required at high levels for high-quality eggs to develop. It is possible this combining of oocyte material occurs in fish and frogs as well, although extensive apoptosis has not been described and in vivo lineage tracing methods would need to be developed to test this idea. Thus, the function of the germ line cyst in vertebrates may depend on the species, serving to increase total oocyte number in aquatic vertebrates or selecting the healthiest oocytes for survival in mammals. Mouse and medaka oocytes can also develop from nonclonal clusters (Saito et al., 2007; Mork et al., 2012), suggesting the germ line cyst is not absolutely required for oogenesis, although the distinction between the two mechanisms is not clear. In this regard, Bb formation may be more robust in fish and frogs to accommodate the lack of intercellular materials from neighboring oocytes. Mouse oocytes form a Bb and are transiently polarized, although the Bb disperses uniformly in the oocyte without conferring polarity to the mature egg. These studies highlight that oocyte development and polarization with the interconnected ovarian cyst is a conserved aspect of animal development and may be highly modified in evolution according to reproductive strategy.

Polarity and the Ovarian Stem Cell

In well-studied invertebrates, Drosophila and *Caenorhabditis elegans*, and in teleost fish, germ line cysts in the ovary derive from ovarian stem cells and these cells continue to divide and to produce oocytes in the adult (Spradling et al., 2011). In teleosts, ovarian stem cells have been identified (medaka: Nakamura et al., 2010; zebrafish: Beer and Draper, 2012); these cells express *nanos* homologs and reside in a gonadal mesoderm niche bounded by *sox9* expression. Additionally, these ovarian stem cells persist into adulthood and are active in oogenesis (Beer and Draper, 2012). Because fish, flies, and worms have polarized oocytes and eggs, it is tempting to hypothesize that robust oocyte polarity and Bb formation might be correlated with the persistence of ovarian stem cells. However, this idea may be too simplistic. Amphibians have long-been thought to replenish oocytes into adulthood, based on high fecundity year-after-year (Duellman and Trueb, 1986; Gilbert, 2011). However, adult oogenesis in amphibians has not been widely studied for the presence or absence of neo-oocytes; such studies would be necessary to confirm or disconfirm the connection between early oocyte polarization and the formation of ovarian stem cells. In line with this idea, a recent report in Rana suggests that frogs may generate their stock of oocytes early in life (Ogielska et al., 2013), as in mammals and most other vertebrates. The idea that oogenesis occurs strictly during fetal development in mammals has recently been controversially disputed however (Dunlop and Telfer, 2014; Hanna and Hennebold, 2014). Additionally, evidence suggests male spermatogonial stem cells arise from completely fragmented cysts in mouse (Lei and Spradling, 2013b), which could suggest that the ability to generate ovarian stem cells could correlate with the extent of cyst fragmentation and proper stem cell niche formation.

STRUCTURAL BASIS FOR MITOCHONDRIAL CLOUD/ BALBIANI FORMATION

Although the Bb has been observed and morphologically described for many years, only recently has its structural components been investigated. Early studies focused on the ultrastructure and cytoskeletal composition (Billett and Adam, 1976; Heasman et al., 1984). Additional work in *Xenopus* showed an accumulation of microtubules and eventually a cytokeratin "net," which is correlated with the Bb becoming resistant to disruption with microtubule-destabilizing drugs (Heasman et al., 1984; Gard, 1999). Other mitochondrial aggregates surrounding the GV do not show this stabilization, suggesting a specific structural alteration. As the common appellation for the Bb suggests, the mitochondrial cloud contains abundant mitochondria along with granular—fibrillar material (nuage), considered the precursor to the germ granules (Heasman et al., 1984). However, the organelles comprising the Bb can vary between species, with Golgi stacks predominating in the mouse Bb (Pepling et al., 2007). Although it is tempting to speculate that these differences are related to the different functions of these bodies in development, there are few functional studies addressing this issue.

While mysterious for many years, recent studies have provided considerable insight to the structural basis for Bb formation and into its role in establishing oocyte polarity. Forward maternal-effect genetic screens in zebrafish identified two mutants with dramatic effects on oocyte polarity: *buckyball* (*buc*) and *magellan/microtubule cross-linking factor 1* (*macf1*), the only such genes implicated in animal—vegetal polarity in vertebrates (Dosch et al., 2004; Wagner et al., 2004; Marlow and Mullins, 2008; Bontems et al., 2009; Gupta et al., 2010). *buc* mutant oocytes lack Bbs, are deficient in vegetal mRNA localization, and show global overall polarity defects including failure of blastodisc reorganization to the animal pole and the formation of ectopic micropyles (Dosch et al., 2004; Marlow and Mullins, 2008). The micropyle is normally a derivative of animal pole follicle cells. It has been shown that loss of Buc from the oocyte is sufficient to mispattern the follicle cells, suggesting oocyte-derived signals are involved. In *macf1* mutants, Bb assembly occurs normally, but vegetal localization and Bb disassembly were deficient, resulting in mispositioning of both the perdurant Bb as well as the GV (Dosch et al., 2004; Gupta et al., 2010).

Macf1

Whereas *buc* encodes an enigmatic protein (see below), the Macf1 protein structure suggested roles in organizing the cytoskeleton during oocyte polarization. Macf1 is related to the Spectraplakin family of proteins, large cytoskeletal scaffolding proteins implicated in various aspects of cell structure and polarity (Bottenberg et al., 2009; Suozzi et al., 2012; Escobar-Aguirre et al., 2017). And as its name suggests, Macf1 is thought to link microtubules to actin and possibly other cytoskeletal elements, such as intermediate filaments (Suozzi et al., 2012), providing a potential structural basis for animal–vegetal polarity. Also, because the Bb is not properly disassembled, Macf1 may play a role in this process, or it may be secondary to the putative structural function. Interestingly, Macf1 is homologous to Drosophila Shortstop, a critical protein required for formation of the fusome and for polarity in oogonial cysts, potentially by regulating microtubule organization and transport (Bottenberg et al., 2009). Macf1 likely does not function analogously in zebrafish, since Bb and polarity defects are seen only after cyst formation (Gupta et al., 2010), but other spectrasome components or related spectraplakin proteins may act in the early cyst.

Buckyball

Buc is essential for the formation of the Bb and for animal–vegetal polarity in the zebrafish oocyte (Dosch et al., 2004; Marlow and Mullins, 2008; Bontems et al., 2009). The Buc protein is conserved in vertebrates and is homologous to Velo1, identified in *Xenopus* as vegetally localized mRNA (Claussen and Pieler, 2004; Bontems et al., 2009). The N-terminus of Buckyball is more highly conserved and is a predicted prion-like domain (Boke et al., 2016), whereas much of the protein is composed of intrinsically disordered regions. These have been implicated in the formation of ribonucleoprotein (RNP) granules by promoting liquid phase–like transitions. Overexpression of *buc* mRNA in zebrafish embryos organizes ectopic germ plasm and induces supernumerary PGCs (Bontems et al., 2009), suggesting a role analogous to *oskar* in Drosophila pole plasm assembly in the egg and embryo. Buc and Velo1 are also implicated in overall Bb assembly. However, although germ plasm originates within the Bb, it is unclear whether Buc/Velo1 is acting similarly in both cases. Recent data from zebrafish and *Xenopus* suggest that Buc/Velo1 plays a central role in the self-organization of the Bb (Heim et al., 2014; Elkouby et al., 2016; Escobar-Aguirre et al., 2017) and functions structurally in Bb assembly by forming an amyloid-like scaffold (Boke et al., 2016; Boke and Mitchison, 2017).

In zebrafish, both antibody and functional GFP fusion transgene studies show that Buc protein is localized to the Bb, along with its mRNA (Heim et al., 2014; Riemer et al., 2015). Heim et al. (2014) showed that Buc is asymmetrically aggregated prior to full Bb assembly, suggesting a role in the early stages of Bb formation. Additionally, this study also found that expression of *buc* from a transgene containing introns was required to rescue oocyte polarity (Heim et al., 2014). Furthermore, analysis of these transgenes in wild-type females (or *buc* heterozygotes) indicated an antimorphic effect, resulting in mislocalization of *buc* RNA and another Bb-localized mRNA, *vasa*, in a manner dependent on functional Buc protein. These data thus suggested a self-organizing mechanism whereby proper localization of *buc* RNA, likely involving splicing-dependent assembly of a localization competent RNP particle, is required for the local production of Buc protein, which further promotes Bb assembly. This idea is supported by the observation that expression of an intronless *buc* transgene with the 3'UTR (likely required for full localization) can induce supernumerary Bbs (Heim et al., 2014), likely the result of generating small foci of Buc synthesis, which activates a feedback loop forming the Bb. Further evidence of positive feedback mechanisms in Bb assembly came from protein interaction data showing that Buc binds to Dazl and Rbpms2 (*alias* Hermes), translation-activating proteins, whereas these proteins bind to *buc* and likely other mRNAs.

Thus, one model for the feedback mechanism is that recruitment of Dazl and Rbpms2 proteins by Buc further recruits *buc* mRNA and establishes local high levels of Buc synthesis in the nascent Bb. It is unclear if this mechanism is fully conserved however, since *velo1* is not Bb-localized in *Xenopus* (Claussen and Pieler, 2004), but it is possible that Velo1 is enriched by other mechanisms. Additional genetic analysis of Buc function in zebrafish showed that initial polarization at the zygotene bouquet stage occurred normally in maternal *buc* mutants (Elkouby et al., 2016). Therefore, regardless of the mechanism of its enrichment, Buc-mediated assembly of the Bb is critical for the elaboration of animal—vegetal polarity downstream of this event.

Interestingly, complementary biochemical studies in *Xenopus* show that the Bb is densely enriched in amyloid-like fibrils and that Velo1 is a prominent component of this amyloid network (Boke et al., 2016). Velo1 protein is recruited to the Bb via its N-terminal prion-like domain (PLD) in both frogs and zebrafish (Nijjar and Woodland, 2013; Heim et al., 2014; Boke et al., 2016) and the purified protein forms amyloid-like fibrils in vitro (Boke et al., 2016). Bb-localized Velo1 has low mobility in fluorescence recovery after photobleaching experiments, suggesting it forms a stable protein network typical of amyloid fibrils. Purified Velo1 protein directly clusters mitochondria as well as nonspecific RNA (Boke et al., 2016). The latter activity requires the C-terminal region, which contains positive K/R-rich motifs. *Xenopus* Velo1 also encodes a shorter variant, lacking the C-terminal IDR (Velo1-Short). This protein is uniformly expressed and is not enriched in the Bb, although it does accumulate in later-stage germ plasm islands (Nijjar and Woodland, 2013). Depletion of this *velo1-short* RNA (but not the full-length RNA) with antisense oligos resulted in dispersal of germ plasm islands in cultured stage VI oocytes (Nijjar and Woodland, 2013), although its role in early Bb assembly was not tested. Curiously, Velo1 may be phosphorylated following Bb disassembly and formation of germ plasm in the egg and likely loses its amyloid properties by this point, based on SDS sensitivity assays (Boke et al., 2016; Boke and Mitchison, 2017). Correspondingly, Velo1 and other proteins are mobile in stage VI germ plasm, which would also be inconsistent with an inert amyloid structure, as suggested for the bulk Bb (Nijjar and Woodland, 2013; Boke et al., 2016). Thus, germ plasm formation may entail a transition out of amyloid into more mobile liquid-like assemblages based on different states of Buc/Velo1 protein networks.

Genetic and other functional data regarding Buc/Velo1 and Bb formation in general in vertebrates are exclusively from zebrafish at this point and work from other models would be informative. The Velo1 gene is predicted to be nonfunctional (as a protein) or pseudogenized in mammals and possibly ax-olotls (Bontems et al., 2009; Škugor et al., 2016). Because oocytes in most organisms, including mammals, do form Bbs (Guraya, 1979; Kloc et al., 2004b, 2008; Pepling et al., 2007)[1], at least transiently, other proteins must be involved in Bb assembly in those cases. Mitochondrial clustering activity has been reported for the *Xenopus* protein Pgat/Xpat (Machado et al., 2005), which has no known homologs. The importance of this activity is unclear; unlike Velo1, Pgat lacks a PLD, but it does contain large intrinsically disordered regions (our unpublished observations). Thus, it is possible that many unique and species-specific proteins may have evolved roles in Bb assembly.

Despite its long history and prominence, the basic function of the Bb has not been conclusively established. Because the Bb is present in oocytes that do not contribute polarized information to the egg, it is likely this aspect is secondary. Also, germ cells typically contain related structures, such as the germ plasm/nuage and sperm chromatoid body, the Bb may have a more general function during oocyte development, such as housing gene products critical for RNA processing/storage and transposon silencing. The linkage of Bb formation to a meiosis-specific event (i.e., zygotene bouquet formation; Elkouby et al., 2016) may thus ensure that this structure only forms in cells undergoing oogenesis. Also, a general requirement for the Bb in oogenesis would allow it to serve as a cryptic "preadaptation" for the repeated evolu-tionary recruitment of germ plasm in the egg of different animals.

GENERATION OF CORTICAL POLARITY IN THE OOCYTE AND EGG

The ability of animal—vegetal polarity of the oocyte/egg to influence early cell fate in some animals has been amply demonstrated. However, animal—vegetal polarity in the egg cortex also has a more general role in determining the site

---

1. Axolotl oocytes are reported to lack a mitochondrial cloud/Bb (Johnson et al., 2001), although it is not known whether this a specific derived condition or whether Bbs form very transiently or cryptically.

of sperm entry. Additionally, the extent that the egg cortex undergoes apical/ basal polarization can differentiate inner from outer cells. This section will outline the mechanisms that establish polarity within the plane of the egg cortex and the consequences of this polarity for fertilization and development.

Cortical Polarity and Sperm Entry

In *Xenopus*, fully grown oocytes are arrested in diplotene of meiosis I and the oocyte is animally−vegetally polarized, with the GV positioned eccentrically toward the animal pole. This positioning occurs during the elaboration of animal−vegetal polarization during oogenesis and during oocyte maturation, resulting in the migration of GVBD contents animally and the asymmetric division and extrusion of the polar body at animal cortex. Classical and more recent embryological experiments showed that sperm entry and fertilization occurred primarily in the animal hemisphere of frog eggs (Newport, 1853; Hertwig, 1877; Roux, 1887; Elinson, 1975). Parenthetically, it was noted by Newport and others that the sperm entry site was largely correlated with the ventral side of the embryo, which lead to the study of the cytoplasmic rearrangements surrounding the gray crescent (reviewed elsewhere (Houston, 2017)). This tendency of animal pole sperm entry is conserved in some lineages, notably in some fish, which form a tunnel in the animal pole follicle cells, the micropyle, facilitating sperm entry (Amanze and Iyengar, 1990; Murata, 2003). The evolution of the micropyle has occurred in some lineages with polarized eggs as an additional redundancy, possibly in response to sperm competition. Micropyle formation is dependent on prior animal−vegetal polarization of the oocyte; *buc* mutant zebrafish oocytes lack Bbs and vegetal identity and correspondingly form ectopic micropyles (Marlow and Mullins, 2008).

However, this situation is far from universal. The eggs of urodeles (newts and salamanders), despite outwardly resembling frog eggs, exhibit obligate polyspermy and permit sperm entry in either animal or vegetal hemisphere (Fankhauser, 1932; Fankhauser and Moore, 1941). The selection of sperm that undergo pronuclear fusion in these organisms is not known, although animal-entering sperm may be favored. In large eggs, it is likely advantageous for pronuclear fusion if the sperm enters animally and is thus in proximity to the female pronucleus, which by definition is in the animal hemisphere.

In unpolarized eggs such as sea urchin and mammals, sperm entry is generally thought to occur anywhere on the egg surface. One exception is the formation of a microvilli-free zone around the site of polar body extrusion in the mammalian egg (Johnson et al., 1975; Nicosia et al., 1977; Longo and Chen, 1985) that is refractive to sperm entry. In this case, GV migration to the cortex occurs stochastically toward one side, subsequently altering the cortex in that region and establishing the presumptive animal pole (Maro et al., 1986; Verlhac et al., 2000). The mechanisms that establish this polarity are thought to be related to spindle attachment at the cortex and involve the apical PAR proteins (Vinot et al., 2004, 2005). Fertilization is biased toward the animal hemisphere in mouse eggs (Motosugi et al., 2006), but this is hypothesized to occur through physical constraints imparted by the zona pellucida. Thus, whereas there may be relocalization of sperm receptors to the animal half in response to local meiotic spindle, this has not been demonstrated and is not strictly necessary for fertilization under experimental conditions.

Apical–Basal Polarization of the *Xenopus* Egg

Despite their animal–vegetal polarization, oocytes do not exhibit functional apical–basal polarity across the animal–vegetal axis. Visible markers such as microvilli (apical) are found uniformly over the oocyte surface (Dumont, 1972; Rasar and Hammes, 2006). Also, proteins typically restricted to apical and basolateral membranes are found in both animal and vegetal regions. These include both the apical marker atypical protein kinase C (Prkci/aPKC/Protein kinase C, iota (Nakaya et al., 2000)) and basolateral proteins Cadherin 3 and Integrin beta 1 (Cdh3; P-cadherin; XB/U-cadherin; and Itgb1 (Angres et al., 1991; Müller et al., 1992, 1993)). Oocyte maturation is thought to establish a presumptive apicobasal polarity superimposed on the animal–vegetal axis, with the animal hemisphere corresponding to apical and the vegetal hemisphere to basolateral (Fig. 1.3). The mechanisms establishing this polarization are thought to rely on the aPKC/PAR system of reciprocal interactions and occur largely independently of overall cytoskeletal changes, since animal–vegetal cytoskeletal differences are eliminated during maturation.

During oocyte maturation, global secretion is blocked and transmembrane proteins, including the Cadherins and Integrins, are internalized into post-Golgi vesicles (Roberts et al., 1992; Müller and Hausen, 1995). Whether this event is correlative or causative, aPKC/PAR components become segregated in the egg cortex, with apical Prkci and Par/partitioning defective homologs 3 and 6 (Pard3 and Pard6) becoming localized to the animal cortex (Choi et al., 2000; Nakaya et al., 2000) and basal Lethal giant larvae homolog 2 (Llgl2) localizing to the vegetal cortex (Cha et al., 2011). In polarized epithelial cells, Prkci is thought to act in a complex with Pard3/6 and in concert with the Crumbs complex to maintain reciprocal localization with the Scribble complex, containing Scrib, Llgl2, and Dlg proteins (St Johnston and Ahringer, 2010). Depletion of *prkci* in *Xenopus* eggs results in expansion of injected Llgl2-GFP localization to the animal cortex (Cha et al., 2011), suggesting that Pkc-Llgl antagonism operates during the oocyte-to-egg transition. These experiments also showed that Prkci interacted with Van Gogh-like 2 (Vangl2), and both were required for microtubule stability in oocytes and for maintaining vegetal Llgl2-GFP and vegetal cortical mRNA localization (of non-germ plasm mRNAs). Interestingly, embryos derived from either *prkci-* or *vangl2*-depleted oocytes are partially ventralized, suggesting that global regulation and coupling of apicobasal and planar cell polarity is critical for

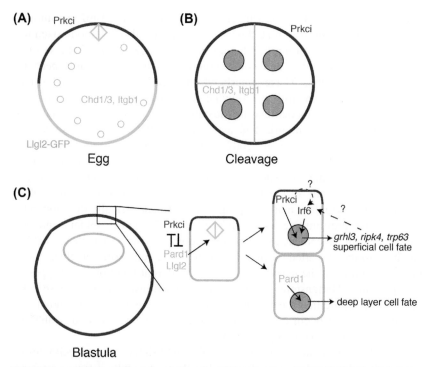

FIGURE 1.3 **Polarity of the egg and the role of the cortex in superficial cell fate.** (A) Polarity is established during oocyte maturation; Prkci is enriched animally (apical, purple), Llgl2 is enriched vegetally (green), and cell surface proteins are internalized (Cdh1/3, Itgb1, blue). This establishes a presumptive apicobasal polarity superimposed over animal–vegetal polarity. (B) During cleavage, cell surface proteins are inserted into newly formed membrane (basolateral, blue), leaving the entire surface of the egg to acquire apical character. (C) In the blastula, cells inheriting egg cortical/apical membrane retain Prkci and differentially localize Pard1 (and other proteins not shown). Pard1 regulates spindle orientation, inducing asymmetric cell division. Cells retaining apical membrane upregulate Irf6 activity, which upregulates target genes and promotes superficial epithelium differentiation. Irf6 target genes likely feedback to regulate Irf6 activity, and Prkci may also influence Irf6 function (*dashed arrows*).

normal axis formation (Cha et al., 2011). These pathways could be required for maintaining *wnt11b* RNA localization or for facilitating the assembly of vegetal cortical microtubules during subsequent cortical rotation. Thus, this transient establishment of apicobasal polarity across the animal–vegetal axis likely has important consequences for both fertilization and for subsequent development of the egg.

Cortical Polarity and Superficial Cell Fate

Apicobasal polarity in the egg is lost after fertilization; the entire egg cortex becomes apical in character, whereas basolateral character arises only in newly generated membrane during cleavage (Roberts et al., 1992; Müller and Hausen, 1995). This was first shown by Holtfreter (1943) in fate mapping experiments, demonstrating that the apical membrane domain from the frog egg is directly inherited and contributes to the apical side of superficial epithelial blastomeres. These superficial outer cells form an epithelial-like barrier to the external milieu, allowing regulated ion and water transport that generate the blastocoel. During the early cleavage stage, superficial cells can undergo both symmetric and asymmetric cell division (Drysdale and Elinson, 1992; Chalmers et al., 2002). For cells dividing within the plane of the epithelium, both daughter cells retain the apical domain and form superficial epithelial cells. These cells will typically differentiate into mucus-secreting goblet cells in the larval epidermis, although in the neural plate they become secondary neuronal progenitors. Asymmetrically dividing cells generate one presumptive superficial cell with the apical domain and one unpolarized inner cell lacking the apical domain. These inner cells differentiate into the so-called "sensorial" layer of the ectoderm, which includes primary neurons in the nervous system, and different functional cell types in the epidermis (Hartenstein, 1989; Drysdale and Elinson, 1992). These cells include the multiciliated cells and ion-secreting cells (*alias* mitochondrial-rich cells), which begin differentiating in the inner layer and subsequently intercalate into the outer superficial layer.

How these apical and basolateral signals establish the different cell fates in polarized and nonpolarized ectodermal cells is not fully known. In both *Xenopus* and zebrafish, superficial cell fate is specified by the maternal transcription factor Irf6 (Interferon regulatory factor 6; Sabel et al., 2009). This gene was identified in human cases of syndromic cleft lip and/or palate (CL/P) associated with Van der Woude (VWS), popliteal pterygium syndromes (PPS) and in cases of nonsyndromic CL/P (Kondo et al., 2002; Kayano et al., 2003; Wang et al., 2003; Ghassibé et al., 2004; Chakravarti, 2004; Zucchero et al., 2004; Leslie et al., 2012, 2013). Irf6 is required for overall epidermal development in mouse, whereas in human cases, haploinsufficiency is observed and defects are likely restricted to the simple epithelium of the oral periderm (Ingraham et al., 2006; Richardson et al., 2006). *Irf6* mRNA is highly maternally expressed in fish and frog eggs but is not localized to the outer superficial layer (Sabel et al., 2009). The protein distribution of Irf6 is not known, but it is likely more active in the superficial layer, although apical signaling regulating Irf6 has not been elucidated. Irf6 regulates a conserved set of genes involved in epidermal development, including *grainyhead-like* genes, which are themselves implicated in simple epithelial development and in human VWS (Chalmers et al., 2006; de la Garza et al., 2013). Interestingly, this superficial epithelium regulatory network also orchestrates epithelial wound healing to some extent, with *Grhl3*-deficient embryos and *Irf6*-deficient mice showing impaired wound healing (Jones et al., 2010; Caddy et al., 2010; Biggs et al., 2014, 2015;

Kousa and Schutte, 2016). This impairment is also seen in human VWS cases, in which there are often more complications following surgical repair of cleft palate (Jones et al., 2010). Recent data have also implicated *ripk4*, implicated in a severe form of PPS, as a downstream target of Irf6 in mice and frogs (De Groote et al., 2015). These studies have suggested a mechanism for control of actin and Cdh1/E-cadherin by Ripk4 function in the superficial epithelium (De Groote et al., 2015). Apical Prkci function is also thought to be critical for superficial fate, mostly through controlling basolateral Pard1 localization (see below). Nuclear Prkci has also been implicated in directly mediating superficial fate in the outer cells (Sabherwal et al., 2009), but it has not been linked to Irf6 function.

In the nonpolarized inner cells, inner cell fate (e.g., primary neuron or ciliated cell) is inhibited by Prkci signaling (Chalmers et al., 2005; Dollar et al., 2005) and promoted by Pard1 activity (Ossipova and Green, 2007; Ossipova et al., 2009). During asymmetric division of outer cells, Prkci activity restricts Pard1 to the basolateral domain through phosphorylation, whereupon Pard1 is thought to promote vertical spindle orientation (Tabler et al., 2010). Subsequent to promoting inner cell fate, Pard1 also regulates Delta–Notch activity, disrupting lateral inhibition and promoting neurogenesis and ciliated cell fates among neural and nonneural inner layer cell ectoderm, respectively (Ossipova and Green, 2007; Ossipova et al., 2009). This Delta–Notch pathway is thought to be the upstream regulator of proneural genes (in the neural plate) and of promulticiliated cells in the epidermis (Deblandre et al., 1999; Stubbs et al., 2006; Quigley et al., 2011). Overall, these observations show that inheritance of polarized apical membrane from the egg promotes superficial fate in cells that retain it following the onset of asymmetric cell divisions through two main mechanisms; (1) inhibition of deep layer fate through Prkci signaling and (2) an Irf6-dependent promotion of superficial fate. These early maternally derived mechanisms are likely reiterated in various contexts in later development, contributing to simple epithelium differentiation in epidermis and oral periderm development.

CONCLUDING REMARKS

In terms of development and disease, defects in oogenesis arising from abnormal polarity would likely contribute to cases of idiopathic infertility. These instances may simply be attributed to "poor-quality" oocytes. Thus, a greater understanding of the underlying polarity of the oocyte during its development would contribute to understanding and identifying potential determinants of oocyte health. Additionally, because of the potential for regenerative/stem cell strategies in reproductive medicine, it is also important to understand the extent of adult oogenesis in mammals and other model organisms. Because the extent of this capacity for mammalian oogenesis remains controversial (Dunlop and Telfer, 2014; Hanna and Hennebold, 2014), studies in model organisms could lead to clues for understanding and activating the process more efficiently.

Also, understanding the mechanisms of oocyte development would better inform efficient directed differentiation of oocytes from pluripotent stem cells (Hayashi et al., 2012; Hikabe et al., 2016) for regenerative medicine. Conversely, the ability to generate oocytes in vitro will likely be critical for further understanding basic mechanisms of germ cell development, including that of humans. Ethical considerations aside however, it has proven difficult to adapt oocyte culture and fertilization methods to the longer timeline of human oogenesis (Smitz and Gilchrist, 2016). However, the mere knowledge that reconstitution of oogenesis from stem cells in vitro is possible will likely induce some to attempt this feat to generate human oocytes.

Studies in model organisms have begun to elucidate the basic structural mechanisms underlying oocyte polarity in detail. Genetic studies in zebrafish have identified roles for Buc/Velo1 and Macf1 in the assembly and disassembly of the Bb, the key determinant of oocyte polarity. Biochemical studies in *Xenopus* have also suggested that the Buc/Velo1 protein likely acts through amyloid network formation. Whereas the role of pathological amyloid aggregation in diseases such as Huntington's and Alzheimer's disease has become apparent (Knowles et al., 2014), it is also evident that regulated amyloid formation is a fundamental property of many cellular structures, including nuclear bodies (e.g., the nucleolus) and the Bb (Hayes and Weeks, 2016; Boke et al., 2016; Boke and Mitchison, 2017). Importantly, it is also becoming increasingly realized that the formation of these bodies and various intracellular granules in general, including the germ granules, relies on the aggregation of RNA-binding proteins through low-complexity domains in liquid-like phase transitions or amyloid-like condensation (Brangwynne et al., 2009, 2011; Kato et al., 2012; Han et al., 2012; Toretsky and Wright, 2014; Courchaine et al., 2016). Thus, investigating these and other processes during oocyte polarity is likely to shed light on many basic cellular and disease-relevant events.

ACKNOWLEDGMENTS

This work was supported by the University of Iowa and by NIH grant GM083999 to DWH.

REFERENCES

Alarcón, V.B., Elinson, R.P., 2001. RNA anchoring in the vegetal cortex of the *Xenopus* oocyte. J. Cell Sci. 114, 1731–1741.

al-Mukhtar, K., Webb, A., 1971. An ultrastructural study of primordial germ cells, oogonia and early oocytes in *Xenopus laevis*. J. Embryol. Exp. Morphol. 26, 195–217.

Amanze, D., Iyengar, A., 1990. The micropyle: a sperm guidance system in teleost fertilization. Development 109, 495–500.

Ancel, P., Vintembenger, P., 1948. Recherches sur le déterminisme de la symétrie bilatérale dans l'œuf des amphibiens. Bull Biol. Fr. Belg. 31, 1−182. Suppl.

Angres, B., Müller, A.H., Kellermann, J., Hausen, P., 1991. Differential expression of two cadherins in *Xenopus laevis*. Development 111, 829−844.

Beer, R.L., Draper, B.W., 2012. nanos3 maintains germline stem cells and expression of the conserved germline stem cell gene nanos2 in the zebrafish ovary. Dev. Biol. https://doi.org/10.1016/j.ydbio.2012.12.003.

Betley, J.N., Heinrich, B., Vernos, I., et al., 2004. Kinesin II mediates Vg1 mRNA transport in *Xenopus* oocytes. Curr. Biol. 14, 219−224. https://doi.org/10.1016/j.cub.2004.01.028.

Biggs, L.C., Naridze, R.L., DeMali, K.A., et al., 2014. Interferon regulatory factor 6 regulates keratinocyte migration. J. Cell Sci. 127, 2840−2848. https://doi.org/10.1242/jcs.139246.

Biggs, L.C., Goudy, S.L., Dunnwald, M., 2015. Palatogenesis and cutaneous repair: a two-headed coin. Dev. Dyn. 244, 289−310. https://doi.org/10.1002/dvdy.24224.

Billett, F.S., Adam, E., 1976. The structure of the mitochondrial cloud of *Xenopus laevis* oocytes. J. Embryol. Exp. Morphol. 36, 697−710.

Boke, E., Mitchison, T.J., 2017. The Balbiani body and the concept of physiological amyloids. Cell Cycle 16, 153−154. https://doi.org/10.1080/15384101.2016.1241605.

Boke, E., Ruer, M., Wühr, M., et al., 2016. Amyloid-like self-assembly of a cellular compartment. Cell 166, 637−650. https://doi.org/10.1016/j.cell.2016.06.051.

Bontems, F., Stein, A., Marlow, F., et al., 2009. Bucky ball organizes germ plasm assembly in zebrafish. Curr. Biol. 19, 414−422. https://doi.org/10.1016/j.cub.2009.01.038.

Bottenberg, W., Sanchez-Soriano, N., Alves-Silva, J., et al., 2009. Context-specific requirements of functional domains of the Spectraplakin short stop in vivo. Mech. Dev. 126, 489−502. https://doi.org/10.1016/j.mod.2009.04.004.

Bounoure, L., 1931. Sur l'existence d'un *déterminant germinal* dans l'oeuf indivis de la Grenouille rousse. CR Acad. Sci. Paris 193, 402.

Bounoure, L., 1934. Recherches sur la lignée germinale chez la grenouille rousse aux premiers stades du développment. Ann. Sci. Nat. 17, 67−248.

Bounoure, L., 1937. Le sort de la lignée germinale chez la Grenouille rousse après l'"action des rayons ultra-violets sur le pôle inférieur de l'"oeuf. CR Acad. Sci. Paris 204, 1837.

Brangwynne, C.P., Eckmann, C.R., Courson, D.S., et al., 2009. Germline P granules are liquid droplets that localize by controlled dissolution/condensation. Science 324, 1729−1732. https://doi.org/10.1126/science.1172046.

Brangwynne, C.P., Mitchison, T.J., Hyman, A.A., 2011. Active liquid-like behavior of nucleoli determines their size and shape in *Xenopus laevis* oocytes. Proc. Natl. Acad. Sci. 108, 4334−4339. https://doi.org/10.1038/nature03207.

Caddy, J., Wilanowski, T., Darido, C., et al., 2010. Epidermal wound repair is regulated by the planar cell polarity signaling pathway. Dev. Cell 19, 138−147. https://doi.org/10.1016/j.devcel.2010.06.008.

Cha, S.-W., Tadjuidje, E., Wylie, C.C., Heasman, J., 2011. The roles of maternal Vangl2 and aPKC in *Xenopus* oocyte and embryo patterning. Development. https://doi.org/10.1242/dev.068866.

Chakravarti, A., 2004. Finding needles in haystacks−IRF6 gene variants in isolated cleft lip or cleft palate. N. Engl. J. Med. 351, 822−824. https://doi.org/10.1056/NEJMe048164.

Chalmers, A.D., Welchman, D., Papalopulu, N., 2002. Intrinsic differences between the superficial and deep layers of the *Xenopus* ectoderm control primary neuronal differentiation. Dev. Cell 2, 171−182.

Chalmers, A.D., Pambos, M., Mason, J., et al., 2005. aPKC, Crumbs3 and Lgl2 control apicobasal polarity in early vertebrate development. Development 132, 977−986. https://doi.org/10.1242/dev.01645.

Chalmers, A.D., Lachani, K., Shin, Y., et al., 2006. Grainyhead-like 3, a transcription factor identified in a microarray screen, promotes the specification of the superficial layer of the embryonic epidermis. Mech. Dev. 123, 702−718. https://doi.org/10.1016/j.mod.2006.04.006.

Chang, P., Torres, J., Lewis, R.A., et al., 2004. Localization of RNAs to the mitochondrial cloud in *Xenopus* oocytes through entrapment and association with endoplasmic reticulum. Mol. Biol. Cell 15, 4669−4681. https://doi.org/10.1091/mbc.E04-03-0265.

Choi, S.-C., Kim, J., Han, J.-K., 2000. Identification and developmental expression of par-6 gene in *Xenopus laevis*. Mech. Dev. 91, 347−350. https://doi.org/10.1016/S0925-4773(99)00281-6.

Ciemerych, M.A., Mesnard, D., Zernicka-Goetz, M., 2000. Animal and vegetal poles of the mouse egg predict the polarity of the embryonic axis, yet are nonessential for development. Development 127, 3467−3474.

Claussen, M., Pieler, T., 2004. Xvelo1 uses a novel 75-nucleotide signal sequence that drives vegetal localization along the late pathway in *Xenopus* oocytes. Dev. Biol. 266, 270−284.

Coggins, L.W., 1973. An ultrastructural and radioautographic study of early oogenesis in the toad *Xenopus laevis*. J. Cell Sci. 12, 71−93.

Conklin, E., 1905. Mosaic development in Ascidian eggs. J. Exp. Zool. 2, 145−223.

Courchaine, E.M., Lu, A., Neugebauer, K.M., 2016. Droplet organelles? Embo J. 35, 1603−1612. https://doi.org/10.15252/embj.201593517.

De Groote, P., Tran, H.T., Fransen, M., et al., 2015. A novel RIPK4-IRF6 connection is required to prevent epithelial fusions characteristic for popliteal pterygium syndromes. Cell Death Differ. 22, 1012−1024. https://doi.org/10.1038/cdd.2014.191.

de la Garza, G., Schleiffarth, J.R., Dunnwald, M., et al., 2013. Interferon regulatory factor 6 promotes differentiation of the periderm by activating expression of grainyhead-like 3. J. Invest. Dermatol. 133, 68−77. https://doi.org/10.1038/jid.2012.269.

Deblandre, G.A., Wettstein, D.A., Koyano-Nakagawa, N., Kintner, C., 1999. A two-step mechanism generates the spacing pattern of the ciliated cells in the skin of *Xenopus* embryos. Development 126, 4715−4728.

Dollar, G.L., Weber, U., Mlodzik, M., Sokol, S.Y., 2005. Regulation of lethal giant larvae by dishevelled. Nature 437, 1376−1380. https://doi.org/10.1038/nature04116.

Dosch, R., Wagner, D.S., Mintzer, K.A., et al., 2004. Maternal control of vertebrate development before the midblastula transition: mutants from the zebrafish I. Dev. Cell 6, 771−780. https://doi.org/10.1016/j.devcel.2004.05.002.

Drysdale, T., Elinson, R., 1992. Cell migration and induction in the development of the surface ectodermal pattern of the *Xenopus* tadpole. Dev. Growth Differ. 34, 51−59.

Duellman, W.E., Trueb, L., 1986. Biology of Amphibians. McGraw-Hill, New York, NY.

Dumont, J.N., 1972. Oogenesis in *Xenopus laevis* (Daudin). I. Stages of oocyte development in laboratory maintained animals. J. Morphol. 136, 153−179. https://doi.org/10.1002/jmor.1051360203.

Dunlop, C.E., Telfer, E.E., 2014. Ovarian germline stem cells. Stem Cell 5, 98−105.

Elinson, R.P., 1975. Site of sperm entry and a cortical contraction associated with egg activation in the frog *Rana pipiens*. Dev. Biol. 47, 257−268.

Elkouby, Y.M., Jamieson-Lucy, A., Mullins, M.C., 2016. Oocyte polarization is coupled to the chromosomal bouquet, a conserved polarized nuclear configuration in meiosis. PLoS Biol. 14, e1002335. https://doi.org/10.1371/journal.pbio.1002335.

Escobar-Aguirre, M., Elkouby, Y.M., Mullins, M.C., 2017. Localization in oogenesis of maternal regulators of embryonic development. Adv. Exp. Med. Biol. 953, 173−207. https://doi.org/10.1007/978-3-319-46095-6_5.

Fankhauser, G., Moore, C., 1941. Cytological and experimental studies of polyspermy in the newt, Triturus viridescens I. Normal fertilization. J. Morphol. 68, 347−385. https://doi.org/10.1002/jmor.1050680208.

Fankhauser, G., 1932. Cytological studies on egg fragments of the salamander triton. II. The history of the supernumerary sperm nuclei in normal fertilization and cleavage of fragments containing the egg nucleus. J. Exp. Zool. 62, 185−235. https://doi.org/10.1002/jez.1400620108.

Forristall, C., Pondel, M., Chen, L., King, M., 1995. Patterns of localization and cytoskeletal association of two vegetally localized RNAs, Vg1 and Xcat-2. Development 121, 201−208.

Gard, D.L., Affleck, D., Error, B.M., 1995. Microtubule organization, acetylation, and nucleation in *Xenopus laevis* oocytes: II. A developmental transition in microtubule organization during early diplotene. Dev. Biol. 168, 189−201. https://doi.org/10.1006/dbio.1995.1071.

Gard, D.L., 1991. Organization, nucleation, and acetylation of microtubules in *Xenopus laevis* oocytes: a study by confocal immunofluorescence microscopy. Dev. Biol. 143, 346−362.

Gard, D.L., 1994. Gamma-tubulin is asymmetrically distributed in the cortex of *Xenopus* oocytes. Dev. Biol. 161, 131−140. https://doi.org/10.1006/dbio.1994.1015.

Gard, D.L., 1999. Confocal microscopy and 3-D reconstruction of the cytoskeleton of *Xenopus* oocytes. Microsc. Res. Tech. 44, 388−414. https://doi.org/10.1002/(SICI)1097-0029(19990315)44:6<388::AID-JEMT2>3.0.CO;2-L.

Gardner, R.L., 1998. Axial relationships between egg and embryo in the mouse. Curr. Top. Dev. Biol. 39, 35−71.

Geigy, R., 1931. Action de l"ultra-violet sur le p61e germinal dans l"oeuf de Drosophila melanogaster (Castration et mutabilit6). Rev. Suisse Zool. 38, 187588.

Ghassibé, M., Revencu, N., Bayet, B., et al., 2004. Six families with van der Woude and/or popliteal pterygium syndrome: all with a mutation in the IRF6 gene. J. Med. Genet. 41, e15.

Gilbert, S.F., 2011. Developmental Biology, tenth ed. Sinauer Associates, Sunderland, MA.

Grieder, N.C., de Cuevas, M., Spradling, A.C., 2000. The fusome organizes the microtubule network during oocyte differentiation in Drosophila. Development 127, 4253−4264.

Gupta, T., Marlow, F.L., Ferriola, D., et al., 2010. Microtubule actin crosslinking factor 1 regulates the Balbiani body and animal-vegetal polarity of the zebrafish oocyte. PLoS Genet. 6, e1001073. https://doi.org/10.1371/journal.pgen.1001073.

Guraya, S.S., 1979. Recent advances in the morphology, cytochemistry, and function of Balbiani's vitelline body in animal oocytes. Int. Rev. Cytol. 59, 249−321.

Han, T.W., Kato, M., Xie, S., et al., 2012. Cell-free formation of RNA granules: bound RNAs identify features and components of cellular assemblies. Cell 149, 768−779. https://doi.org/10.1016/j.cell.2012.04.016.

Hanna, C.B., Hennebold, J.D., 2014. Ovarian germline stem cells: an unlimited source of oocytes? Fertil. Steril. 101, 20−30. https://doi.org/10.1016/j.fertnstert.2013.11.009.

Harper, L., Golubovskaya, I., Cande, W.Z., 2004. A bouquet of chromosomes. J. Cell Sci. 117, 4025−4032. https://doi.org/10.1242/jcs.01363.

Hartenstein, V., 1989. Early neurogenesis in *Xenopus*: the spatio-temporal pattern of proliferation and cell lineages in the embryonic spinal cord. Neuron 3, 399−411.

Hayashi, K., Ogushi, S., Kurimoto, K., Shimamoto, S., Ohta, H., Saitou, M., 2012. Offspring from oocytes derived from in vitro primordial germ cell-like cells in mice. Science 338, 971−975. https://doi.org/10.1126/science.1226889.

Hayes, M.H., Weeks, D.L., 2016. Amyloids assemble as part of recognizable structures during oogenesis in *Xenopus*. Biol. Open. https://doi.org/10.1242/bio.017384.

Heasman, J., Quarmby, J., Wylie, C.C., 1984. The mitochondrial cloud of *Xenopus* oocytes: the source of germinal granule material. Dev. Biol. 105, 458−469.

Hegner, R.W., 1914. Studies on germ cells. I. The history of the germ cells in insects with special reference to the Keimbahn-determinants. II. The origin and significance of the Keimbahn-determinants in animals. J. Morphol. 25, 375−509.

Heim, A.E., Hartung, O., Rothhämel, S., et al., 2014. Oocyte polarity requires a Bucky ball-dependent feedback amplification loop. Development 141, 842−854. https://doi.org/10.1242/dev.090449.

Hertig, A.T., 1968. The primary human oocyte: some observations on the fine structure of Balbiani's vitelline body and the origin of the annulate lamellae. Am. J. Anat. 122, 107−137. https://doi.org/10.1002/aja.1001220107.

Hertwig, O., 1877. Beiträge zur Kenntniss der Bildung, Befruchtung und Theilung des thierischen Eis. Morphol. Jahrb. 3, 1−86.

Hiiragi, T., Solter, D., 2004. First cleavage plane of the mouse egg is not predetermined but defined by the topology of the two apposing pronuclei. Nature 430, 360−364. https://doi.org/10.1038/nature02595.

Hikabe, O., Hamazaki, N., Nagamatsu, G., Obata, Y., Hirao, Y., Hamada, N., Shimamoto, S., Imamura, T., Nakashima, K., Saitou, M., Hayashi, K., 2016. Reconstitution in vitro of the entire cycle of the mouse female germ line. Nature 539, 299−303. https://doi.org/10.1038/nature20104.

Holtfreter, J., 1943. Properties and functions of the surface coat in amphibian embryos. J. Exp. Zool. 93, 251−323. https://doi.org/10.1002/jez.1400930205.

Houston, D.W., King, M.L., 2000. Germ plasm and molecular determinants of germ cell fate. Curr. Top. Dev. Biol. 50, 155−181.

Houston, D.W., 2013. Regulation of cell polarity and RNA localization in vertebrate oocytes. Int. Rev. Cell Mol. Biol. 306, 127−185. https://doi.org/10.1016/B978-0-12-407694-5.00004-3.

Houston, D.W., 2017. Vertebrate axial patterning: from egg to asymmetry. Adv. Exp. Med. Biol. 953, 209−306. https://doi.org/10.1007/978-3-319-46095-6_6.

Howley, C., Ho, R.K., 2000. mRNA localization patterns in zebrafish oocytes. Mech. Dev. 92, 305−309. https://doi.org/10.1016/S0925-4773(00)00247-1.

Huettner, A.F., 1923. The origin of the germ cells in Drosophila melanogaster. J. Morphol. 37, 385−423. https://doi.org/10.1002/jmor.1050370204.

Illmensee, K., Mahowald, A.P., 1974. Transplantation of posterior polar plasm in Drosophila. Induction of germ cells at the anterior pole of the egg. PNAS 71, 1016−1020.

Ingraham, C., Kinoshita, A., Kondo, S., et al., 2006. Abnormal skin, limb and craniofacial morphogenesis in mice deficient for interferon regulatory factor 6 (Irf6). Nat. Genet. 38, 1335−1340.

Johnson, M.H., Eager, D., Muggleton-Harris, A., Grave, H.M., 1975. Mosaicism in organisation concanavalin A receptors on surface membrane of mouse egg. Nature 257, 321−322.

Johnson, A.D., Bachvarova, R.F., Drum, M., Masi, T., 2001. Expression of axolotl DAZL RNA, a marker of germ plasm: widespread maternal RNA and onset of expression in germ cells approaching the gonad. Dev. Biol. 234, 402–415. https://doi.org/10.1006/dbio.2001.0264.

Jones, J., Canady, J.W., Brookes, J.T., 2010. Wound complications following cleft repair in children with Van der Woude syndrome.

Kato, M., Han, T.W., Xie, S., et al., 2012. Cell-free formation of RNA granules: low complexity sequence domains form dynamic fibers within hydrogels. Cell 149, 753–767. https://doi.org/10.1016/j.cell.2012.04.017.

Kayano, S., Kure, S., Suzuki, Y., et al., 2003. Novel IRF6 mutations in Japanese patients with Van der Woude syndrome: two missense mutations (R45Q and P396S) and a 17-kb deletion.

King, M.L., Messitt, T.J., Mowry, K.L., 2005. Putting RNAs in the right place at the right time: RNA localization in the frog oocyte. Biol. Cell 97, 19–33. https://doi.org/10.1042/BC20040067.

Kloc, M., Etkin, L.D., 1995. Two distinct pathways for the localization of RNAs at the vegetal cortex in *Xenopus* oocytes. Development 121, 287–297.

Kloc, M., Larabell, C., Etkin, L.D., 1996. Elaboration of the messenger transport organizer pathway for localization of RNA to the vegetal cortex of *Xenopus* oocytes. Dev. Biol. 180, 119–130. https://doi.org/10.1006/dbio.1996.0289.

Kloc, M., Zearfoss, N.R., Etkin, L.D., 2002. Mechanisms of subcellular mRNA localization. Cell 108, 533–544.

Kloc, M., Bilinski, S., Dougherty, M.T., et al., 2004a. Formation, architecture and polarity of female germline cyst in *Xenopus*. Dev. Biol. 266, 43–61.

Kloc, M., Bilinski, S., Etkin, L.D., 2004b. The Balbiani body and germ cell determinants: 150 years later. Curr. Top. Dev. Biol. 59, 1–36. https://doi.org/10.1016/S0070-2153(04)59001-4.

Kloc, M., Jaglarz, M., Dougherty, M., et al., 2008. Mouse early oocytes are transiently polar: three-dimensional and ultrastructural analysis. Exp. Cell Res. 314, 3245–3254. https://doi.org/10.1016/j.yexcr.2008.07.007.

Klymkowsky, M.W., Maynell, L.A., Polson, A.G., 1987. Polar asymmetry in the organization of the cortical cytokeratin system of *Xenopus laevis* oocytes and embryos. Development 100, 543–557.

Knowles, T.P.J., Vendruscolo, M., Dobson, C.M., 2014. The amyloid state and its association with protein misfolding diseases. Nat. Rev. Mol. Cell Biol. 15, 384–396. https://doi.org/10.1038/nrm3810.

Kondo, S., Schutte, B.C., Richardson, R.J., Bjork, B.C., 2002. Mutations in IRF6 cause Van der Woude and popliteal pterygium syndromes.

Kousa, Y.A., Schutte, B.C., 2016. Toward an orofacial gene regulatory network. Dev. Dyn. 245, 220–232. https://doi.org/10.1002/dvdy.24341.

Lei, L., Spradling, A.C., 2013a. Female mice lack adult germ-line stem cells but sustain oogenesis using stable primordial follicles. Proc. Natl. Acad. Sci. 110, 8585–8590. https://doi.org/10.1073/pnas.1306189110.

Lei, L., Spradling, A.C., 2013b. Mouse primordial germ cells produce cysts that partially fragment prior to meiosis. Development 140, 2075–2081. https://doi.org/10.1242/dev.093864.

Lei, L., Spradling, A.C., 2016. Mouse oocytes differentiate through organelle enrichment from sister cyst germ cells. Science 352, 95–99. https://doi.org/10.1126/science.aad2156.

Lenhart, K.F., DiNardo, S., 2015. Somatic cell encystment promotes abscission in germline stem cells following a regulated block in cytokinesis. Dev. Cell 34, 192–205. https://doi.org/10.1016/j.devcel.2015.05.003.

Leslie, E.J., Standley, J., Compton, J., Bale, S., 2012. Comparative analysis of IRF6 variants in families with Van der Woude syndrome and popliteal pterygium syndrome using public whole-exome databases.

Leslie, E.J., Standley, J., Compton, J., et al., 2013. Comparative analysis of IRF6 variants in families with Van der Woude syndrome and popliteal pterygium syndrome using public whole-exome databases. Genet. Med. 15, 338−344. https://doi.org/10.1038/gim.2012.141.

Longo, F.J., Chen, D.Y., 1985. Development of cortical polarity in mouse eggs: involvement of the meiotic apparatus.

Machado, R.J., Moore, W., Hames, R., et al., 2005. *Xenopus* Xpat protein is a major component of germ plasm and may function in its organisation and positioning. Dev. Biol. 287, 289−300. https://doi.org/10.1016/j.ydbio.2005.08.044.

Marlow, F.L., Mullins, M.C., 2008. Bucky ball functions in Balbiani body assembly and animal-vegetal polarity in the oocyte and follicle cell layer in zebrafish. Dev. Biol. 321, 40−50. https://doi.org/10.1016/j.ydbio.2008.05.557.

Maro, B., Johnson, M.H., Webb, M., Flach, G., 1986. Mechanism of polar body formation in the mouse oocyte: an interaction between the chromosomes, the cytoskeleton and the plasma membrane. J. Embryol. Exp. Morphol. 92, 11−32.

Messitt, T.J., Gagnon, J.A., Kreiling, J.A., et al., 2008. Multiple kinesin motors coordinate cyto-plasmic RNA transport on a subpopulation of microtubules in *Xenopus* oocytes. Dev. Cell 15, 426−436. https://doi.org/10.1016/j.devcel.2008.06.014.

Miyamoto, K., Pasque, V., Jullien, J., Gurdon, J.B., 2011. Nuclear actin polymerization is required for transcriptional reprogramming of Oct4 by oocytes. Genes Dev. 25, 946−958. https://doi.org/10.1101/gad.615211.

Miyamoto, K., Teperek, M., Yusa, K., et al., 2013. Nuclear Wave1 is required for reprogramming transcription in oocytes and for normal development. Science 341, 1002−1005. https://doi.org/10.1126/science.1240376.

Mork, L., Tang, H., Batchvarov, I., Capel, B., 2012. Mouse germ cell clusters form by aggregation as well as clonal divisions. Mech. Dev. 128, 591−596. https://doi.org/10.1016/j.mod.2011.12.005.

Motosugi, N., Dietrich, J.-E., Polanski, Z., et al., 2006. Space asymmetry directs preferential sperm entry in the absence of polarity in the mouse oocyte. PLoS Biol. 4, e135. https://doi.org/10.1371/journal.pbio.0040135.

Müller, H.A., Hausen, P., 1995. Epithelial cell polarity in early *Xenopus* development. Dev. Dyn. 202, 405−420. https://doi.org/10.1002/aja.1002020410.

Müller, A.H., Angres, B., Hausen, P., 1992. U-cadherin in *Xenopus* oogenesis and oocyte matu-ration. Development 114, 533−543.

Müller, A.H., Gawantka, V., Ding, X., Hausen, P., 1993. Maturation induced internalization of beta 1-integrin by *Xenopus* oocytes and formation of the maternal integrin pool. Mech. Dev. 42, 77−88.

Murata, K., 2003. Blocks to polyspermy in fish: a brief review. Aquaculture and pathobiology of crustacean and other species. In: Sakai, Y. (Ed.), Aquaculture and Pathobiology of Crustacean and Other Species: Proceedings of the Thirty-second U.S. Japan Symposium on Aquaculture. NOAA Research.

Nakamura, S., Kobayashi, K., Nishimura, T., et al., 2010. Identification of germline stem cells in the ovary of the teleost medaka. Science 328, 1561−1563. https://doi.org/10.1126/science.1185473.

Nakaya, M., Fukui, A., Izumi, Y., et al., 2000. Meiotic maturation induces animal-vegetal asymmetric distribution of aPKC and ASIP/PAR-3 in *Xenopus* oocytes. Development 127, 5021−5031.

Newport, G., 1853. On the Impregnation of the Ovum in the Amphibia. (Second Series, Revised.) And on the Direct Agency of the Spermatozoon.

Nicosia, S.V., Wolf, D.P., Inoue, M., 1977. Cortical granule distribution and cell surface characteristics in mouse eggs. Dev. Biol. 57, 56–74.

Nieuwkoop, P.D., Sutasurya, L.A., 1976. Embryological evidence for a possible polyphyletic origin of the recent amphibians. J. Embryol. Exp. Morphol. 35, 159–167.

Nieuwkoop, P.D., 1969. The formation of mesoderm in Urodelean amphibians. I. Induction by the endoderm. Wilhelm Roux Arch. Entwickl. Mech. Org. 162, 341–373.

Nijjar, S., Woodland, H.R., 2013. Protein interactions in Xenopus germ plasm RNP particles.

Ogielska, M., Kotusz, A., Augustyńska, R., et al., 2013. A stockpile of ova in the grass frog *Rana temporaria* is established once for the life span. Do ovaries in amphibians and in mammals follow the same evolutionary strategy? Anat. Rec. (Hoboken) 296, 638–653. https://doi.org/10.1002/ar.22674.

Ossipova, T., Green, S., 2007. PAR1 specifies ciliated cells in vertebrate ectoderm downstream of aPKC. Development 134, 4297–4306. https://doi.org/10.1242/dev.009282.

Ossipova, O., Ezan, J., Sokol, S.Y., 2009. PAR-1 phosphorylates mind bomb to promote vertebrate neurogenesis. Dev. Cell 17, 222–233. https://doi.org/10.1016/j.devcel.2009.06.010.

Pepling, M.E., Spradling, A.C., 2001. Mouse ovarian germ cell cysts undergo programmed breakdown to form primordial follicles. Dev. Biol. 234, 339–351. https://doi.org/10.1006/dbio.2001.0269.

Pepling, M.E., de Cuevas, M., Spradling, A.C., 1999. Germline cysts: a conserved phase of germ cell development? Trends Cell Biol. 9, 257–262.

Pepling, M.E., Wilhelm, J.E., O'Hara, A.L., et al., 2007. Mouse oocytes within germ cell cysts and primordial follicles contain a Balbiani body. Proc. Natl. Acad. Sci. 104, 187–192. https://doi.org/10.1073/pnas.0609923104.

Pepling, M.E., 2006. From primordial germ cell to primordial follicle: mammalian female germ cell development. Genesis 44, 622–632. https://doi.org/10.1002/dvg.20258.

Pfeiffer, D.C., Gard, D.L., 1999. Microtubules in *Xenopus* oocytes are oriented with their minus-ends towards the cortex. Cell Motil. Cytoskelet. 44, 34–43. https://doi.org/10.1002/(SICI)1097-0169(199909)44:1<34::AID-CM3>3.0.CO;2–6.

Quigley, I.K., Stubbs, J.L., Kintner, C., 2011. Specification of ion transport cells in the *Xenopus* larval skin. Development 138, 705–714. https://doi.org/10.1242/dev.055699.

Rasar, M.A., Hammes, S.R., 2006. The physiology of the *Xenopus laevis* ovary. Methods Mol. Biol. 322, 17–30. https://doi.org/10.1007/978-1-59745-000-3_2.

Richardson, R., Dixon, J., Malhotra, S., et al., 2006. Irf6 is a key determinant of the keratinocyte proliferation-differentiation switch. Nat. Genet. 38, 1329–1334.

Riemer, S., Bontems, F., Krishnakumar, P., et al., 2015. A functional Bucky ball-GFP transgene visualizes germ plasm in living zebrafish. Gene Expr. Patterns 18, 44–52. https://doi.org/10.1016/j.gep.2015.05.003.

Roberts, S.J., Leaf, D.S., Moore, H.P., Gerhart, J.C., 1992. The establishment of polarized membrane traffic in *Xenopus laevis* embryos. J. Cell Biol. 118, 1359–1369.

Rossant, J., Tam, P.P.L., 2009. Blastocyst lineage formation, early embryonic asymmetries and axis patterning in the mouse. Development 136, 701–713. https://doi.org/10.1242/dev.017178.

Roux, W., 1887. Beiträge zur Entwickelungsmechanik des Embryo. Arch. Mikrosk. Anat. 29, 157–211.

Sabel, J.L., d'Alençon, C., O'Brien, E.K., et al., 2009. Maternal interferon regulatory factor 6 is required for the differentiation of primary superficial epithelia in Danio and *Xenopus* embryos. Dev. Biol. 325, 249–262. https://doi.org/10.1016/j.ydbio.2008.10.031.

Sabherwal, N., Tsutsui, A., Hodge, S., et al., 2009. The apicobasal polarity kinase aPKC functions as a nuclear determinant and regulates cell proliferation and fate during *Xenopus* primary neurogenesis. Development 136, 2767—2777. https://doi.org/10.1242/dev.034454.

Saito, D., Morinaga, C., Aoki, Y., et al., 2007. Proliferation of germ cells during gonadal sex differentiation in medaka: insights from germ cell-depleted mutant zenzai. Cell 310, 280—290. https://doi.org/10.1016/j.ydbio.2007.07.039.

Sardet, C., Prodon, F., Dumollard, R., et al., 2002. Structure and function of the egg cortex from oogenesis through fertilization. Dev. Biol. 241, 1—23. https://doi.org/10.1006/dbio.2001.0474.

Selman, K., Wallace, R.A., Sarka, A., Qi, X., 1993. Stages of oocyte development in the zebrafish, Brachydanio rerio. J. Morphol. 218, 203—224.

Škugor, A., Tveiten, H., Johnsen, H., Andersen, Ø., 2016. Multiplicity of Buc copies in Atlantic salmon contrasts with loss of the germ cell determinant in primates, rodents and axolotl. BMC Evol. Biol. 16, 232. https://doi.org/10.1186/s12862-016-0809-7.

Smitz, J.E.J., Gilchrist, R.B., 2016. Are human oocytes from stem cells next? Nat. Biotechnol. 34, 1247—1248. https://doi.org/10.1038/nbt.3742.

Spradling, A., Fuller, M.T., Braun, R.E., Yoshida, S., 2011. Germline stem cells. Cold Spring Harb. Perspect. Biol. 3, a002642. https://doi.org/10.1101/cshperspect.a002642.

St Johnston, D., Ahringer, J., 2010. Cell polarity in eggs and epithelia: parallels and diversity. Cell 141, 757—774. https://doi.org/10.1016/j.cell.2010.05.011.

Stubbs, J.L., Davidson, L., Keller, R., Kintner, C., 2006. Radial intercalation of ciliated cells during *Xenopus* skin development. Development 133, 2507—2515. https://doi.org/10.1242/dev.02417.

Sudarwati, S., Nieuwkoop, P.D., 1971. Mesoderm formation in the anuran *Xenopus laevis* (Daudin). Dev. Genes Evol. 166, 189—204. https://doi.org/10.1007/BF00650029.

Suozzi, K.C., Wu, X., Fuchs, E., 2012. Spectraplakins: master orchestrators of cytoskeletal dynamics. J. Cell Biol. 197, 465—475. https://doi.org/10.1083/jcb.201112034.

Sutasurya, L.A., Nieuwkoop, P.D., 1974. The induction of the primordial germ cells in the Urodeles. Wilhelm Roux Arch. Entwickl. Mech. Org. 175, 199—220.

Tabler, J.M., Yamanaka, H., Green, J.B.A., 2010. PAR-1 promotes primary neurogenesis and asymmetric cell divisions via control of spindle orientation. Development 137, 2501—2505. https://doi.org/10.1242/dev.049833.

Tada, H., Mochii, M., Orii, H., Watanabe, K., 2012. Ectopic formation of primordial germ cells by transplantation of the germ plasm: direct evidence for germ cell determinant in *Xenopus*. Dev. Biol. 371, 86—93. https://doi.org/10.1016/j.ydbio.2012.08.014.

Toretsky, J.A., Wright, P.E., 2014. Assemblages: functional units formed by cellular phase separation. J. Cell Biol. 206, 579—588. https://doi.org/10.1083/jcb.201404124.

Van Blerkom, J., Bell, H., 1986. Regulation of development in the fully grown mouse oocyte: chromosome-mediated temporal and spatial differentiation of the cytoplasm and plasma membrane. J. Embryol. Exp. Morphol. 93, 213—238.

Verlhac, M.H., Lefebvre, C., Guillaud, P., et al., 2000. Asymmetric division in mouse oocytes: with or without Mos. Curr. Biol. 10, 1303—1306.

VerMilyea, M.D., Maneck, M., Yoshida, N., et al., 2011. Transcriptome asymmetry within mouse zygotes but not between early embryonic sister blastomeres. Embo J. 30, 1841—1851. https://doi.org/10.1038/emboj.2011.92.

Vinot, S., Le, T., Maro, B., Louvet-Vallée, S., 2004. Two PAR6 proteins become asymmetrically localized during establishment of polarity in mouse oocytes. Curr. Biol. 14, 520—525. https://doi.org/10.1016/j.cub.2004.02.061.

Vinot, S., Le, T., Ohno, S., et al., 2005. Asymmetric distribution of PAR proteins in the mouse embryo begins at the 8-cell stage during compaction. Dev. Biol. 282, 307−319. https://doi.org/10.1016/j.ydbio.2005.03.001.

Wagner, D.S., Dosch, R., Mintzer, K.A., et al., 2004. Maternal control of development at the midblastula transition and beyond: mutants from the zebrafish II. Dev. Cell 6, 781−790. https://doi.org/10.1016/j.devcel.2004.04.001.

Wang, X., Liu, J., Zhang, H., Xiao, M., Li, J., Yang, C., Lin, X., Wu, Z., Hu, L., Kong, X., 2003. Novel mutations in the IRF6 gene for Van der Woude syndrome. Hum. Genet. 113, 382−386. https://doi.org/10.1007/s00439-003-0989-2.

Wilson, E., 1928. The Cell in Development and Heredity, third ed. The Macmillian Company, New York.

Wylie, C.C., Brown, D., Godsave, S.F., et al., 1985. The cytoskeleton of *Xenopus* oocytes and its role in development. J. Embryol. Exp. Morphol. 89 (Suppl.), 1−15.

Yisraeli, J.K., Sokol, S., Melton, D.A., 1990. A two-step model for the localization of maternal mRNA in *Xenopus* oocytes: involvement of microtubules and microfilaments in the translocation and anchoring of Vg1 mRNA. Development 108, 289−298.

Yoon, Y.J., Mowry, K.L., 2004. *Xenopus* Staufen is a component of a ribonucleoprotein complex containing Vg1 RNA and kinesin. Development 131, 3035−3045. https://doi.org/10.1242/dev.01170.

Zernicka-Goetz, M., Morris, S.A., Bruce, A.W., 2009. Making a firm decision: multifaceted regulation of cell fate in the early mouse embryo. Nat. Rev. Genet. 10, 467−477. https://doi.org/10.1038/nrg2564.

Zhou, Y., King, M.L., 1996. Localization of Xcat-2 RNA, a putative germ plasm component, to the mitochondrial cloud in *Xenopus* stage I oocytes. Development 122, 2947−2953.

Zickler, D., Kleckner, N., 1998. The leptotene-zygotene transition of meiosis. Annu. Rev. Genet. 32, 619−697. https://doi.org/10.1146/annurev.genet.32.1.619.

Zucchero, T.M., Cooper, M.E., Maher, B.S., et al., 2004. Interferon regulatory factor 6 (IRF6) gene variants and the risk of isolated cleft lip or palate. N. Engl. J. Med. 351, 769−780. https://doi.org/10.1056/NEJMoa032909.

Chapter 2

Pluripotency—What Does Cell Polarity Have to Do With It?

Tristan Frum, Amy Ralston
Michigan State University, East Lansing, MI, United States

MAMMALIAN PLURIPOTENCY

Pluripotency is best defined as the ability to produce any of the more than two hundred differentiated cell types in the body. During development, pluripotency is established in epiblast cells. The mission of the pluripotent epiblast is to proliferate and progressively differentiate, producing the mature cell types of the fully formed individual. Thus, pluripotency is a property that is transient during animal development (Fig. 2.1A). Notably, pluripotency is not equivalent to totipotency. This is because pluripotent cells are more restricted in their developmental potential than totipotent cells. The distinction between pluripotent and totipotent is important in animals, such as amniotes, which produce extraembryonic tissues (e.g., yolk sac or placenta) during development. In amniotes, the cells of the embryo are totipotent until the point at which the extraembryonic lineages are first established. At this point, the remaining pluripotent cells, which are destined to produce the animal itself, are, by definition, more limited in their developmental potential because they have lost the ability to produce the extraembryonic tissues. Therefore, understanding how pluripotency is first established means focusing on the transition from totipotent to pluripotent cell fates during embryogenesis.

A common misconception is that pluripotent epiblast cells are those that preserve the totipotent properties of the zygote. This view is inaccurate because, as explained above, the zygote has broader developmental potential than do the epiblast cells. Therefore, a more accurate description is that totipotency is lost during embryogenesis and is replaced by programs that actively drive cells to ensuing specializations, such as pluripotency and extraembryonic lineage formation. Another common misconception is that pluripotency is a permanent or long-lasting state within the embryo. This view is inaccurate because cells of the embryo quickly differentiate to more specialized progenitor pools. For example, pluripotent cells begin differentiation as they

Cell Polarity in Development and Disease. https://doi.org/10.1016/B978-0-12-802438-6.00002-4

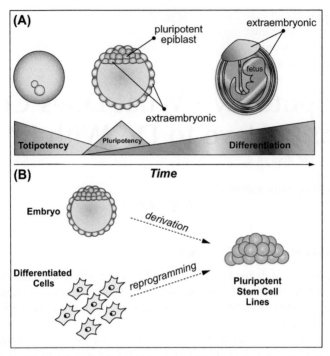

FIGURE 2.1 States of mammalian pluripotency. (A) The mammalian zygote is totipotent because it has the potential to form the fetus and the extraembryonic membranes. During the first few days of development, totipotency is replaced by pluripotency in the epiblast, and multipotency within the extraembryonic lineages. Pluripotent cells can be observed for a few short days of gestation and exit pluripotency at gastrulation, when cells differentiate to form the germ layers. (B) Pluripotent stem cells can be derived from embryos and created by reprogramming somatic cells. Pluripotent stem cells maintain proliferative ability and pluripotency by repressing differentiation.

adopt germ layer fates at the time of gastrulation (E6.5) (Lawson and Pedersen, 1992; Rivera-Pérez and Hadjantonakis, 2014; Tam and Behringer, 1997). Around the time of gastrulation, primordial germ cells (PGCs) are specified de novo, rather than by segregation of predetermined cells (Johnson and Alberio, 2015). While mouse PGCs can give rise to pluripotent embryonic germ cells in vitro (Labosky et al., 1994; Matsui et al., 1992; Resnick et al., 1992), PGCs do not demonstrate pluripotent properties in vivo (Leitch et al., 2014) and are considered to possess latent pluripotency (Leitch and Smith, 2013). Therefore, pluripotency is a transient property of epiblast cells, in that it is actively established, but not maintained, during embryogenesis.

Maintaining the state of pluripotency is possible by deriving embryonic stem cell lines (ESCs) from the embryos of species, such as rodents and primates (Weinberger et al., 2016) (Fig. 2.1B). A common misconception is that pluripotent stem cells are present within the embryo, and that these same cells are propagated in culture as stem cell lines. A more accurate view is that pluripotent stem cell lines are derived from embryos by a process that artificially preserves their capacity to proliferate while pausing their propensity to differentiate. Thus pluripotent stem cell lines differ fundamentally from embryonic cells, consistent with molecular changes observed within pluripotent epiblast cells during derivation of stem cell lines (Boroviak et al., 2014). Notably, many pathways important for suspending differentiation in pluripotent stem cell lines are also involved in diapause—a period of suspended animation that can be induced in some mammalian species (Renfree, 2015). Therefore, while there are ways to pause differentiation in stem cell lines and in embryos, the goal of embryonic cells undergoing active development is to differentiate.

Pluripotency can also be artificially induced in differentiated (somatic) cells to reverse development and create induced pluripotent stem cell lines (iPSCs) (Takahashi et al., 2007; Takahashi and Yamanaka, 2006). This kind of cellular reprogramming provides a unique opportunity to identify the mechanisms *establishing* pluripotency, while pluripotent stem cell lines provide the opportunity to discover the mechanisms *maintaining* pluripotency. The existence of pluripotent stem cell lines, as well as reprogramming technology thus provides a unique way to learn about the molecular regulation and the cell biology of pluripotency.

Cell polarity is important for regulating a number of important cellular events, many of which are important for stem cell and developmental biology. Given the many ways that cell polarity can affect cellular function and gene expression, it is important to evaluate how cell polarity impacts the establishment and maintenance of pluripotency. This chapter will review what is currently known about the role of cell polarity in the context of three pluripotency paradigms: the mouse embryo, embryo-derived stem cell lines, and somatic cell reprogramming.

CELL POLARITY IN THE MAMMALIAN EMBRYO

Spoiler: In the early embryo, cell polarization is repressed in emerging pluripotent epiblast cells, while apicobasal polarization is established in nonpluripotent cells. In the established pluripotent epiblast, cells polarize just prior to exiting pluripotency. Therefore, cell polarization is a hallmark of the loss of pluripotency.

The Emergence of Pluripotency in the Mouse Embryo

Cell fate specification has been especially well studied in mice because of the easy access to embryos and an arsenal of genetic tools. More recently, human embryos undergoing the first cell fate decisions have been scrutinized, and many similarities in morphogenesis and regulated gene expression have been noted (Chazaud and Yamanaka, 2016), underscoring the utility of the mouse to understand early mammalian development and the origins of pluripotency. In considering how the first cell fate decisions are regulated during mammalian development, genetic studies of mouse embryogenesis have provided several lessons. First, embryonic cell fates are plastic and embryonic cell polarity is labile (Frum and Ralston, 2015). Second, patterning and polarizing the oocyte does not play an important role in early mammalian development as it does in nonmammalian species (Nance, 2014). Rather, cell fates are emergent properties of the early embryo. Third, cell polarization is a hallmark of a more differentiated state and is repressed during the establishment pluripotency (reviewed here). Thus, pluripotent cells are the unpolarized cells within the early embryo, while the nonpluripotent cells of the extraembryonic lineages are the polarized ones. Later in development, the maturing epiblast becomes polarized, just prior to exit from pluripotency. These observations suggest that cell polarization must be actively repressed to facilitate pluripotency or, alternatively, cell polarization could help drive differentiation.

In mice, as well as other amniote species, embryogenesis involves generating not only the pluripotent epiblast but also the extraembryonic tissues (e.g., placenta and yolk sac), which support and instruct embryonic development. These extraembryonic tissues are so fundamental to embryogenesis that they are the first cell types specified during embryogenesis (Fig. 2.1A). One such extraembryonic lineage, the trophectoderm, facilitates implantation of the embryo into the mother's uterus and, after implantation, will contribute cells to the developing placenta. Another extraembryonic lineage, the primitive endoderm, participates in nourishing the implanting embryo and, after implantation, provides signals that instruct head, heart, and blood formation within the epiblast (Baron et al., 2012; Srinivas, 2006). Concomitant with specification of trophectoderm and primitive endoderm, pluripotent epiblast fate is also first defined. By the blastocyst stage, the epiblast and primitive endoderm form the inner cell mass of a cystlike structure, while the trophectoderm encloses the inner cell mass and a fluid-filled blastocoel as a spherical, polarized epithelium. Because the formation of extraembryonic and pluripotent cell fates are linked temporally and functionally, understanding how pluripotency originates also relies on understanding how the extraembryonic lineages are established or when totipotency is lost.

To evaluate when totipotency is lost and when pluripotency first emerges in the embryo, a number of criteria can be used. First, cellular potency can be defined in terms of developmental potential. To evaluate developmental potential, cells are typically removed from their native environment and transplanted to a new context, where their potential to give rise to one or more cell types can be evaluated. This approach asks what a cell *can* do, but it does not ask what a cell normally *does* do. Second, cellular potency can be defined in terms of fate. To evaluate cell fate, cells are tracked within their native environment to determine which cell types they will eventually produce. This is achieved by genetic or time-lapse lineage tracing. This approach asks what a cell normally *does* do but does not ask what that cell *could* do, given the chance. Third, cellular potency can be defined in terms of genes, where gene expression marks specific cell types, and where gene function defines specific cell types. A variety of approaches are used for evaluating gene expression and gene function in mice (Lokken and Ralston, 2016). All of these approaches have been used to determine when totipotency declines and when and where pluripotency emerges in the mouse early embryo. Not surprisingly, the results of these diverse investigations do not necessarily agree, and this is due, to some extent, to the fact that each approach tests different questions about the potential of the early blastomeres.

Tracking cellular potency in terms of *developmental potential*, embryonic blastomeres demonstrate totipotent properties until the 16-cell stage (E3.0) and, rarely, the 32-cell stage (E3.25) (Suwińska et al., 2008; Tarkowski et al., 2010). Therefore, *some* cells of the embryo are developmentally plastic for the first 2−3 days after fertilization. Still, other studies have shown that some of the cells of the early embryo are less plastic than others, starting as early as the 4-cell stage (Torres-Padilla et al., 2007). Thus, totipotency, defined by the behavior of isolated blastomeres placed in an artificial environment, is lost from some cells of the embryo between the 4-cell and 16-cell stages (E2.0−E3.0). Transplantation experiments have also been used to evaluate when inner cell mass cells become committed to pluripotency. These experiments showed that inner cell mass cells can adopt either primitive endoderm or epiblast fate until the late blastocyst stage (E4.5) (Chazaud et al., 2006; Gardner, 1982, 1984; Gardner and Rossant, 1979; Grabarek et al., 2012; Meilhac et al., 2009). Therefore, a subset of cells within the inner cell mass becomes irreversibly committed to pluripotency by the time of implantation (E4.5), while retaining broader developmental potential prior to implantation. An inclusive model would show that totipotency is progressively and asynchronously lost among blastomeres during preimplantation (Fig. 2.1A).

Tracking cellular potency in terms of *cell fate*, lineage tracing experiments have shown that inner cell mass cells are produced during the fourth, fifth, and sixth cleavages of development, as blastomere cell numbers increase from 8 to 64 cells (McDole and Zheng, 2012; Morris et al., 2010; Pedersen et al., 1986). Inner cell mass cells can be daughters of outside or inside cells since cleavage patterns are variable in mammals. However, the timing of when the inner cell mass cells descend from outside cells may influence their fate (Morris et al., 2010). Another study did not observe any bias (Yamanaka et al., 2010), possibly because cells were transplanted rather than examined in their native context. Alternatively, the number of primary inside and secondary inside cells has been proposed to influence cell fate and signaling (Krupa et al., 2014). Regardless, these observations indicate that pluripotent cells emerge between the 16- and 64-cell stages (E3.0−E3.75).

Tracking pluripotency on the basis of *pluripotency genes* has shown that many genes considered to be reliable markers of pluripotency are not, actually, expressed specifically in pluripotent cells during all stages of development. For instance, several pluripotency genes are initially expressed in totipotent blastomeres, and then their expression becomes progressively restricted to emerging epiblast cells. Two examples are *Nanog* and *Oct4* (*Pou5f1*), which are expressed ubiquitously from the 8-cell stage or earlier, until the blastocyst stage (Mitsui et al., 2003; Palmieri et al., 1994; Strumpf et al., 2005). Later (E3.75−E4.0), the expression of both *Oct4* and *Nanog* is repressed within the trophectoderm by the transcription factor CDX2, thereby establishing the inner cell mass-specific expression of *Oct4* and *Nanog* (Strumpf et al., 2005). However, *Oct4* and *Nanog* exhibit different expression patterns within the inner cell mass starting around E3.75. OCT4 is expressed in all inner cell mass cells, both pluripotent epiblast and differentiating primitive endoderm (Guo et al., 2010; Palmieri et al., 1994). Consistent with its broad expression pattern, OCT4 has been shown to promote primitive endoderm cell fate directly (Frum et al., 2013; Le Bin et al., 2014), in addition to its better-known role as an essential pluripotency gene (Nichols et al., 1998; Niwa et al., 2000). While *Oct4* remains expressed within all inner cell mass cells, *Nanog* is repressed within primitive endoderm cells by the transcription factor GATA6, acting downstream of fibroblast growth factor (FGF) signaling (Chazaud et al., 2006; Kang et al., 2013; Schrode et al., 2014). Thus NANOG is detected exclusively within epiblast cells, and epiblast and primitive endoderm cells can be resolved on the basis of their salt and pepper expression of epiblast and primitive endoderm genes (Fig. 2.2A). How FGF signaling is restricted to nonpluripotent, primitive endoderm cells within the inner cell mass is an unsolved mystery.

Intriguing insight into the initiation of pluripotency has been provided by examining the expression dynamics of *Sox2*. Like OCT4, SOX2 plays important roles in pluripotency and reprogramming (Ralston and Rossant, 2010). However, both the expression dynamics of *Sox2* and the regulation of *Sox2* expression are unique in the preimplantation embryo. *Sox2* is expressed exclusively in inside cells of the embryo at the 16-cell stage (Guo et al., 2010). Moreover, SOX2 is only detected in about half of the inside cells at the 16-cell stage (Wicklow et al., 2014), suggesting that *Sox2* is a specific marker of nascent pluripotency. Interestingly, CDX2 does not restrict SOX2 expression to inside cells, indicating that the regulation of *Sox2* differs from that of *Oct4* and *Nanog*. If SOX2 is the earliest and most specific marker of pluripotency, this raises the question as to whether establishing SOX2 is functionally important for establishing or defining pluripotency. Embryos lacking maternal and zygotic *Sox2* still form epiblast cells, which express *Oct4* and *Nanog* (Wicklow et al., 2014), arguing that *Sox2* is not essential for establishing pluripotency. However, ES cell lines cannot be derived from *Sox2* null embryos (Avilion et al., 2003). Therefore, SOX2 may play a relatively late role in the pluripotency pathway. Nevertheless, the initial *Sox2* expression pattern suggests that pluripotent cells are molecularly distinct from other blastomeres

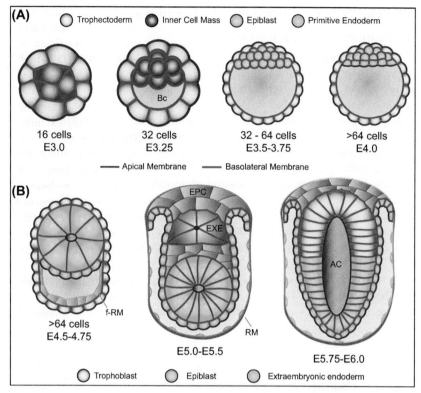

FIGURE 2.2   Cell polarization in the periimplantation mouse embryo. (A) When the embryo reaches 32 cells, blastocyst formation occurs. Cells of the inner cell mass then go on to establish the epiblast and primitive endoderm lineages. Epiblast and primitive endoderm cells emerge intermixed within the inner cell mass in a "salt and pepper" pattern and then rearrange positions. Primitive endoderm cells polarize and form an epithelium after localizing to the blastocoel cavity (*BC*, blastocoel cavity). (B) Around the time of implantation, the epiblast forms a rosette and polarizes with the apical domain located in the center of the rosette. A subset of trophectoderm and primitive endoderm cells migrates away from the epiblast to form Reichert's membrane (*f-RM*, future Reichert's membrane). The remaining trophectoderm cells migrate away from the epiblast and polarize to form the extraembryonic ectoderm (EXE), and the lumen of the EXE and epiblast expand and meet the proamniotic cavity is formed, (*AC*, proamniotic cavity; *RM*, Reichert's membrane).

starting around the 16-cell stage, which is just after the emergence of cell polarity within the embryo (see below). Thereafter, *Sox2* colocalizes with *Nanog* in nascent epiblast cells (Wicklow et al., 2014). Subsequently, the embryo undergoes implantation into the mother's uterus and undergoes massive morphological rearrangements that convert the ball of epiblast cells to a rosette and then to the so-called egg cylinder (Fig. 2.2B). The pluripotent epiblast commences differentiation around E6.5, at the onset of gastrulation.

Cell Polarity in the Mammalian Early Embryo

Fertilization to the Early 8-Cell Stage

The mouse embryo has served as a model to understand the de novo polarization of a protoepithelium since the 1970s. The mouse embryo presents unique opportunities and challenges as a model system to understand the establishment of cell polarity (Box 2.1). After fertilization, the zygote undergoes cleavage division to reach the two-cell stage. Up until this point, development is under control of mRNA and protein produced from the maternal genome during oogenesis. As the 2-cell stage proceeds, maternally provided mRNA is degraded and the zygote's own genome becomes transcriptionally active, transitioning control of development from the maternal to the zygotic genome (Schultz, 1993). As the zygotic genome awakens, slow, asynchronous cleavage divisions ensue, generating additional totipotent blastomeres. At this point, and until the early 8-cell stage, individual blastomeres are morphologically indistinguishable and individually totipotent (Tarkowski and Wróblewska, 1967).

At the early 8-cell stage, blastomeres lack tight and adherens junctions (Ducibella et al., 1975; Ducibella and Anderson, 1975; Magnuson et al., 1977) and are therefore not considered polarized. However, several cytoskeletal

---

BOX 2.1 De Novo Cell Polarization in the Mouse Preimplantation Embryo—Challenges and Opportunities

As a model of mammalian development, the mouse embryo provides exceptional advantages. As a model to study the onset of polarization in a protoepithelium, there are advantages and disadvantages of using the mouse preimplantation embryo. Awareness of the disadvantages is important for appreciating which challenges may have slowed the discovery of mechanisms regulating the emergence of cell polarity in mammalian development. Nevertheless, technological advances in imaging, gene expression analysis, and genetics will continue to help overcome these challenges.

Advantages	Disadvantages
• Relatively few cells	• Genetic redundancy among polarity
• Slow cell cycle	gene paralogues
• Optically transparent embryos	• Genetic redundancy between maternal
• Can be cultured ex vivo for	and zygotic genomes
several days	• Spherical configuration of blastomeres
• Live imaging possible	• Dynamic cellular rearrangements
	• Nonstereotyped cleavage patterns
	• Asynchronous cleavage

proteins are asymmetrically distributed within blastomeres at the early 8-cell stage, including actin, myosin, and microtubules, which localize to the outside, exposed surface of the blastomeres (Houliston et al., 1987; Johnson and Maro, 1984; Sobel, 1983b). Also at the early 8-cell stage and prior, E-cadherin (also known as uvomorulin but referred to as CDH1 hereafter) accumulates at the points of cell—cell contact (Johnson et al., 1986). Notably, cellular asymmetries can be redistributed by changing the areas of contact between and among blastomeres (Johnson and Ziomek, 1981a,b; Sobel, 1983a), indicating that the initial molecular asymmetries are defined by the points of cell contact, and are not inherited from, or determined by, a prior developmental stage. As the 8-cell stage proceeds, the early molecular asymmetries are reinforced, as blastomeres overtly polarize during the process of embryo compaction.

Late 8-Cell Stage: Compaction and the Establishment of Blastomere Polarity

During the late 8-cell stage, blastomeres become more overtly polarized (starting at E2.5), during a process known as compaction (Eckert et al., 2015; Fleming, 1992). Prior to compaction, blastomeres are spherical, contact each other relatively minimally, and the embryo contains relatively large intercellular spaces. By contrast, during compaction blastomeres become more conical, they contact each other more extensively, and intercellular spaces are eliminated from the embryo (Fig. 2.3A). Compaction is dependent on *Cdh1* since compaction is completely disrupted in embryos lacking maternal and zygotic *Cdh1* (Stephenson et al., 2010). Whether other adhesion molecules participate in embryo compaction is not known.

Coincident with compaction, blastomeres begin to show signs of radial polarization (Nance, 2014), or polarization along the radial axis of the embryo. For example, components of the PAR complex, including PARD6B, PAR3, and atypical protein kinases C (aPKC), as well as the ERM family protein EZRIN localize to membrane regions devoid of cell—cell contact, forming apical domain facing the outer surface of the embryo (Eckert et al., 2004; Louvet et al., 1996; Pauken and Capco, 2000; Plusa et al., 2005; Vinot et al., 2005). Meanwhile, at areas of cell—cell contact, tight junctions begin to assemble (Ducibella and Anderson, 1975; Fleming et al., 1989; Magnuson et al., 1977; Pratt, 1985). In addition, adherens junctions, which include β-CATENIN (CTNNB1) and CDH1, assemble basolaterally (De Vries et al., 2004; Magnuson et al.; Ohsugi et al., 1996). Other typically basolaterally localized proteins, including PAR1 (EMK1) and SCRIBBLED, also become localized to regions of cell—cell contact (Tao et al., 2012; Vinot et al., 2005). Therefore, by the time the 8-cell embryo has completed the process of compaction, each blastomere contains a molecularly distinct apical domain at the embryo's surface, and a molecularly distinct basolateral domain at the sites

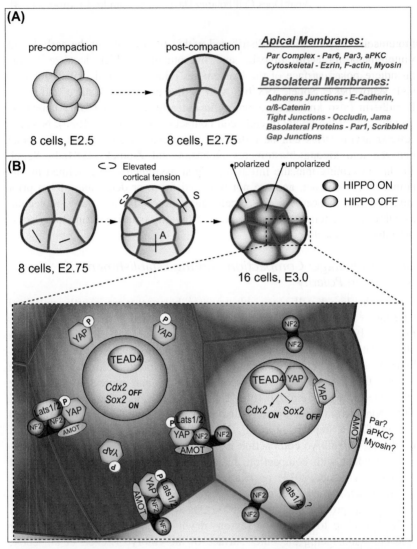

FIGURE 2.3 Cell polarization and regulation of cell fate in the early preimplantation mouse embryo. (A) Coincident with embryo compaction, apicobasal polarization is established in the preimplantation mouse embryo. Basolateral domains are localized to the interior of the embryo, and apical domains face the external environment. (B) After the establishment of apicobasal polarization, inside cells are generated during the next cleavage division. Inside cells are generated by both oriented asymmetric cell divisions and internalization of unpolarized daughter cells resulting from nonoriented asymmetric cell divisions. (*A*, asymmetric cell division; *line*, plane of cell division; *S*, symmetric cell division; *semicircled arrows*, site of cortical tension). In inside cells, HIPPO signaling is active and forms a complex in an Angiomotin (AMOT)-dependent manner to phosphorylate yes-associated protein (YAP), leaving YAP trapped in the cytoplasm. In outside cells, the apical domain sequesters AMOT through a PAR complex-dependent mechanism, which inhibits formation of the HIPPO signaling complex, and leaves YAP unphosphorylated and free to translocate to the nucleus, where YAP partners with TEAD4 to promote trophectoderm and repress inner cell mass gene expression. Contractility also promotes YAP localization to the nucleus, but the precise mechanism is still unclear. Lats1/2 localization in polarized cells is currently unknown.

of cell—cell contact inside the embryo. Given the initial localization of CDH1 to future basolateral regions (Fig. 2.3A), CDH1 has been proposed to organize blastomere polarization at compaction (Hyafil et al., 1980; Peyriéras et al., 1983; Vestweber et al., 1987). Consistent with this prediction, the localization of apical and basolateral proteins becomes overlapping and unpolarized in *Cdh1* null embryos (Stephenson et al., 2010). Recently, CDH1 has also been shown to localize to filopodia observed on the apical surface of the compacting embryo (Fierro-González et al., 2013). Laser ablation of filopodia disrupted cell shape during compaction (ibid.), suggesting they may be important for compaction. However, deleting *Cdh1* exclusively within filopodia is not straightforward, making it difficult to know that filopodia-specific localization of CDH1 is important.

Understanding what triggers compaction is important for understanding the molecular mechanisms regulating the onset of blastomere polarization, but compaction triggers are still unknown. Curiously, compaction not only proceeds, but is accelerated, when protein synthesis is inhibited (Kidder and McLachlin, 1985; Levy et al., 1986), suggesting that the timing of compaction is regulated posttranslationally. Consistent with this prediction, premature activation of aPKC accelerates compaction (Winkel et al., 1990), suggesting that phosphorylation limits the onset of compaction. Notably, premature compaction in aPKC-inhibited embryos is dependent on *Cdh1* (ibid). Therefore, CDH1 activity or localization may be regulated by aPKC phosphorylation. Whether CDH1 is an aPKC substrate in the mouse embryo is unknown. In addition, aPKC is present in blastomeres leading up to the time of compaction (Pauken and Capco, 2000), indicating that additional factors regulate the timing of aPKC activation and the onset of compaction.

The 16-Cell to 32-Cell Stages and the Establishment of Trophectoderm and Inner Cell Mass

The fourth cleavage creates the first positional and polarity differences among blastomeres within the embryo. During the fourth cleavage, polarized 8-cell stage blastomeres undergo asynchronous cleavage divisions to produce the 16-cell embryo, often referred to as the morula. As a consequence of the fourth cleavage division, some cells become positioned within the interior of the embryo (Fig. 2.3B). Meanwhile, apical domain proteins and CDH1 become uniformly localized to the entire cell membranes of inside cells (Ralston and Rossant, 2008; Vinot et al., 2005). Thus inside cells are unpolarized and, because they maintain high levels of CDH1, could be considered to comprise an unpolarized epithelium. By contrast, outside cells maintain their polarized phenotype. Whether cell position drives changes in cell polarization or vice versa is an exciting and ongoing research topic.

The establishment of unpolarized, inside cells and polarized, outside cells coincides with the first cell fate decision in development: trophectoderm versus inner cell mass. The outer, polarized cells not only will give rise to trophectoderm, but can also give rise to more inner cell mass cells, as described above. Around the 32-cell stage, trophectoderm beings to express Na^+/K^+ ATPases and aquaporins (Barcroft et al., 2003; DiZio and Tasca, 1977; Fleming et al., 1989; Watson and Kidder, 1988) and commences functioning as a transporting epithelium. Trophectoderm differentiation drives cavitation by swelling the blastocoel to produce the expanded blastocyst. Thus, during the first cell fate decision, the purpose of cell polarization is to select noninner cell mass cells and prepare them for differentiation.

The 32- to 64-Cell Stages and the Establishment of Epiblast and Primitive Endoderm

After blastocyst formation, the cells of the inner cell mass are progressively sorted into epiblast and primitive endoderm lineages. We have reviewed current models and mechanisms regulating this, the second cell fate decision, elsewhere (Frum and Ralston, 2015; Lokken and Ralston, 2016), and there is no evidence that differential polarization of inner cell mass cells is involved in the second cell fate decision. For example, inner cell mass cells are unpolarized (Vinot et al., 2005) when pluripotent epiblast and primitive endoderm cell fates are first established (E3.75). While epiblast and primitive endoderm fates are first established independent of cell position, in a random, salt and pepper distribution within the inner cell mass, these two cell types subsequently sort into discrete layers by E4.5, by both active cell migration and apoptosis of mislocalized cells (Meilhac et al., 2009; Plusa et al., 2008).

Well after the second cell fate decision has been initiated, around the time of implantation (E.4.5), primitive endoderm cells begin to polarize, evidenced by apical (blastocoelic) localization of aPKC (Saiz et al., 2013). In addition, the low-density lipoprotein receptor LRP2 is expressed specifically in primitive endoderm and localizes to the apical surface of primitive endoderm cells (Gerbe et al., 2008), and the endocytic adaptor protein DAB2, which interacts with LRP2, is also localized to the apical surface of primitive endoderm cells at this stage (Yang et al., 2002). Conversely, Laminins and β1 Integrin (ITGB1) comprise a basement membrane between primitive endoderm and epiblast cells (Moore et al., 2014; Smyth et al., 1999). Therefore, the only cell types to become polarized during the first cell fate decisions are the non-pluripotent extraembryonic lineages, while the pluripotent epiblast cells remain unpolarized until after implantation.

After the 64-Cell Stage: Implantation and Polarization of the Pluripotent Epiblast

Starting at the 64-cell stage, and during the next 24 h, the blastocyst will hatch from its protein coat known as the zona pellucida and implant within the uterine lining. Morphological changes occurring during implantation have not been well studied because the embryo is difficult to access and culture during implantation stages. Recently, however, new culture conditions have been established that facilitate culturing mouse and human embryos through implantation stages (Bedzhov and Zernicka-Goetz, 2014; Deglincerti et al., 2016; Shahbazi et al., 2016).

A recent study focused on characterizing the dynamics of epiblast polarization in mouse embryos (Fig. 2.2B). While the epiblast is unpolarized initially, starting at E4.5, apical (aPKC, PAR6, actin, and myosin) proteins begin to accumulate in the center of the inner cell mass (Bedzhov and Zernicka-Goetz, 2014). Subsequently, epiblast cells become wedge-shaped and arrange into a rosette, with inward-facing apical domains. A lumen that will become the embryo's proamniotic cavity then forms in the center of the rosette—a process proposed to rely on an actomyosin contractions. Curiously, CDH1 initially accumulates in the nascent apical region at E4.5, before settling into a more basal position by E5.5. Polarization of the human epiblast proceeds by a similar series of steps, although CDH1 localization was not examined (Shahbazi et al., 2016). The mechanism proposed to initiate and drive the polarization and cavitation of the postimplantation mouse epiblast is ITGB1 signaling in the epiblast, stimulated by laminin secreted by neighboring primitive endoderm and trophectoderm cells (ibid). One prediction of this model is that the extraembryonic lineages are essential for triggering the polarization of the epiblast. However, the authors did not examine whether epiblast epithelialization fails in the absence of extraembryonic tissues or similar cues provided by the culture milieu. Another prediction of the model is that *Itgb1* is required in epiblast for epiblast cell polarization. Consistent with this, the epiblast is severely disorganized in embryos in which *Itgb1* is deleted within the epiblast (Moore et al., 2014), suggestive of disrupted polarity, although polarity protein distribution was not examined. Subsequent to polarization, the epiblast epithelium undergoes a localized epithelial to mesenchymal transition (EMT) during gastrulation, as the epiblast cells differentiate to the three germ layers of the embryo.

How Cell Polarity Regulates Establishment of Pluripotency in the Mouse Embryo

The previous section described how cell polarity is successively established within cells of the three lineages of the preimplantation embryo and highlighted the fact that polarization of the epiblast occurs well after the establishment of pluripotent epiblast cells. Therefore, cell polarization does not directly establish pluripotency in the mouse early embryo. Nevertheless, cell polarization plays an important indirect role in establishing pluripotency in the early mouse embryo since polarization precedes the first cell fate decision, which is when inner cell mass cells are first selected. The following section will review current models of how cell polarity could regulate the first cell fate decision.

Polarity Regulates Cell Position Within the Embryo

Due to the correlation between cell position, cell polarization, and cell fate, researchers have attempted to identify the causal connection between cell position and polarization. It has long been accepted that inside cells are generated during 2–3 rounds of asymmetric divisions that begin at the 8-cell stage and lead to unequal inheritance of apical membrane components between daughter cells (Fleming, 1987; Johnson and Ziomek, 1981b; Morris et al., 2010). These observations led to the "cell-polarity" model, which proposed that inherited apical domain dictates cell position and cell fate. For example, cleavage planes that transect the apical domain would be symmetric, producing two polarized daughters. However, cleavage planes that do not transect the apical domain would be asymmetric and produce one polarized and one unpolarized cell (Fig. 2.3B). In this model, inheritance of membrane components and cell position is coupled, with the angle of cell division providing the organizing force for proper arrangement of inside unpolarized and outside polar cells. Alternatively, the "inside–outside model" proposed that cell position precedes and causes differences in cell polarization and fate (Suwińska et al., 2008; Tarkowski and Wróblewska, 1967).

Several groups have used time-lapse imaging approaches to examine the correlation between the angle of cell division, cell polarization, and eventual cell fate (indicated by position and/or gene expression). These investigations have provided several important insights that support the cell polarity model, with minor modification. First, rather than dividing perpendicular or parallel to the embryo surface, divisions at the 8-cell and 16-cell stage are often oblique to the embryo surface (McDole et al., 2011; Sutherland et al., 1990; Yamanaka et al., 2010). Second, the angle of cell division does not absolutely correlate with cell fate (Watanabe et al., 2014), suggesting the angle of cell division is insufficient to organize the embryo. Finally, rather than cell position being fixed after asymmetric cell divisions, cells are often internalized after cell division (Plusa et al., 2005; Watanabe et al., 2014; Yamanaka et al., 2010).

Intriguingly, internalized cells often lack an apical domain prior to internalization (Anani et al., 2014), suggesting that polarization, and not the orientation of the cleavage plane itself, drives position. Furthermore, embryos in which either aPKC function is reduced or *Par6d* is knocked down possess more outside cells and fewer inside cells (Plusa et al., 2005), supporting a general role in the activity of apical membrane components in controlling cell position. These observations have led to the prevailing model in which asymmetric inheritance of apical domains influences cell position and cell fate.

Since cell polarization may, in some cases, precede cell position, several groups have focused on investigating the mechanics of how cells become internalized. Recently, a role for differences in cortical tension among blastomeres has been proposed to drive cell internalization. Live imaging of cells undergoing internalization suggested that cells reach the inside region as their membranes undergo apical constriction (Samarage et al., 2015). In addition, direct measurement of relative cortical tension within outside cells and cells undergoing internalization, within living embryos, revealed increased cortical tension in cells during internalization (Maître et al., 2015). Furthermore, elevated cortical tension is predictive of inside cell position and correlates with cell depolarization (Maître et al., 2016), suggesting that cortical tension drives internalization of unpolarized cells. Interestingly, increased cortical tension is observed in the absence of *Cdh1*, consistent with the fact that *Cdh1* null embryos can still produce inside cells. CDH1 has been proposed to focus the region of contractility away from the regions of cell—cell contact, thereby orienting the direction of cell movement toward the inside of the embryo. Therefore, if cell polarization is upstream of cell position, some of the mechanisms translating polarization to position are now becoming clearer.

Polarity Regulates Gene Expression

The correlation between cell polarization and cell fate raises the question of whether cell polarization is instructive of cell fate or, conversely, whether polarization is a consequence of cell fate. Although there is evidence to support both possibilities, it is clear that manipulation of cell polarity leads to dramatic changes in expression of cell fate markers, while manipulating expression of lineage markers leads to subtle or no changes in cell polarization. For example, siRNA-mediated knockdown of *Pard6b* or *aPKC* leads to major defects in trophectoderm cell fate, including failure to express *Cdx2* and to form tight junctions (Alarcon, 2010; Dard et al., 2009). In contrast, cells are able to polarize initially in the absence of either *Cdx2* or its upstream transcriptional activator TEAD4 (Mihajlović et al., 2015; Nishioka et al., 2008; Ralston and Rossant, 2008), demonstrating that cell polarization is upstream of trophectoderm cell fate. Additionally, live imaging of a fluorescent *Cdx2* reporter revealed that the *Cdx2* expression level within blastomeres dynamically adapts to positional information, rather CDX2 driving cell position (McDole and Zheng, 2012; Toyooka et al., 2016). These observations are consistent with the fact that *Cdx2* null blastomeres show no preference for inside or outside position when aggregated with wild-type blastomeres (Ralston and Rossant, 2008). Taken together, these observations argue that cell polarization acts upstream of both cell fate and position.

Some evidence suggests that cell polarization also acts downstream of cell fate. For example, blastomeres injected with *Cdx2* mRNA exhibit increases in apically localized aPKC and preferentially contribute to the trophectoderm lineage (Jedrusik et al., 2008). In addition, the level, but not the localization, of aPKC expression appears moderately diminished in later stage *Cdx2* null embryos, relative to wild-type embryos, suggesting that CDX2 maintains *aPKC* expression after blastocyst formation (Ralston and Rossant, 2008). Taken together, these observations suggest a model in which cell polarity and position act upstream of the expression of *Cdx2*, and CDX2 then reinforces cell fate, including expression of genes encoding cell polarity components, after trophectoderm and inner cell mass cell fate specification.

Cell Polarity Regulates Gene Expression by Repressing HIPPO Signaling

The evidence that cell polarity acts upstream of cell fate has led to a search for factors that pattern trophectoderm and inner cell mass gene expression in response to cell polarization. Research from several different groups has converged on members of the HIPPO signaling pathway as important intermediates between cell polarity and trophectoderm and inner cell mass gene expression. In this section, we discuss how the HIPPO pathway bridges cell polarity and cell fate.

The HIPPO signaling pathway was first identified in *Drosophila* as an important signaling pathway controlling diverse roles in organogenesis, tissue homeostasis, and stem cells (Halder and Johnson, 2011; Tremblay and Camargo, 2012). In the context of the mouse embryo (Fig. 2.3B), one of the main roles of HIPPO signaling is promote the cytoplasmic localization of yes-associated protein (YAP), a transcriptional coactivator that partners with the TEAD-domain containing transcription factor TEAD4 to promote expression of trophectoderm genes such as *Cdx2* and *Gata3* and to repress expression of inner cell mass genes such as *Sox2* (Nishioka et al., 2009; Ralston et al., 2010; Wicklow et al., 2014). Diverse upstream regulators of HIPPO signaling have been identified in a variety of biological contexts, including G protein coupled receptor signaling, cell adhesion, mechanical force, and cell polarity (Moya and Halder, 2016). However, in the mouse early embryo, several lines of evidence indicate that cell polarity is a negative regulator of HIPPO signaling. For example, YAP is localized to nuclei of polarized cells but is phosphorylated and cytoplasmically localized in unpolarized cells of the mouse early embryo (Nishioka et al., 2009). Although cell position and cell polarity are tightly linked in the preimplantation mouse embryo, YAP is cytoplasmic in outside, unpolarized cells that have not yet been internalized, arguing that cell polarity, rather than cell position is the more critical regulator of HIPPO signaling (Anani et al., 2014; Hirate et al., 2015).

Functional evidence that cell polarity represses HIPPO signaling comes from inhibition of Rho-associated kinase (ROCK) activity in early mouse embryos. In ROCK-inhibited (ROCKi) embryos, apical and basolateral components are expressed but are mislocalized along the entire membrane of all cells (Kono et al., 2014). ROCKi-induced disruption of cell polarity leads to exclusion of YAP from the nucleus, disrupting expression of *Cdx2* and trophectoderm specification. Interestingly, nuclear YAP localization can be rescued in ROCKi embryos by knocking down *Lats1/2* or *Angiomotin* (*Amot*) (Kono et al., 2014; Mihajlović and Bruce, 2016), HIPPO signaling components that normally promote the cytoplasmic localization of YAP. Therefore, cell polarity is likely to control YAP localization by repressing the activity of HIPPO signaling components, such as LATS1/2 and AMOT, in outside cells.

The precise mechanistic details of how apical membrane components repress HIPPO signaling are still an active area of investigation. Some insight has been provided by evaluating the localization of HIPPO signaling components within the early mouse embryo. Neurofibromin 2 (NF2), also known as Merlin, is localized in the plasma membrane of all cells of the embryo, indicating that NF2 localization is not regulated by cell polarity (Cockburn et al., 2013). In contrast, the distribution of AMOT is apical in outside cells (Hirate et al., 2013; Leung and Zernicka-Goetz, 2013), and AMOT polarization is dependent on aPKC or PARD6B activity (Hirate et al., 2013). AMOT therefore provides the link between cell polarization and HIPPO signaling because the active HIPPO signaling complex is thought to include AMOT, NF2, LATS1/2, and CDH1 in adherens junctions (Hirate et al., 2013). These observations have led to a model in which AMOT is sequestered at the apical membrane by components of the PAR complex, rendering the basolaterally localized HIPPO complex inactive in polarized cells. In polarized cells lacking active HIPPO signaling, YAP can enter the nucleus and partner with TEAD4 to regulate expression of trophectoderm and inner cell mass genes. In contrast, in inside cells, which lack a discrete apical domain, AMOT associates with the active HIPPO complex, enabling LATS1/2 to phosphorylate YAP, which then remains cytoplasmic.

Recently, cell contractility has also been implicated as a negative regulator of HIPPO signaling. Treatment of embryos with blebbistatin, a selective inhibitor of myosin II, leads to the widespread cytoplasmic localization of YAP, suggesting that actomyosin-mediated cell contractility normally promotes nuclear localization of YAP and represses HIPPO signaling (Maître et al., 2016). However, cell polarization was not examined in blebbistatin-treated embryos, making it unclear whether cortical tension acts independently of cell polarization to regulate YAP localization. Thus, the prevailing model is that HIPPO interprets cell polarity to regulate expression of lineage markers in the mouse blastocyst.

CELL POLARITY AND EMBRYO-DERIVED PLURIPOTENT STEM CELL LINES

Spoiler: As in the embryo, multiple pluripotent states exist among stem cell lines, ranging from naïve to primed, along a pluripotency spectrum. These stem cell states can also be achieved by reprogramming. Among stem cell lines, cell polarization is associated with primed pluripotency, and loss of polarization leads to naïve pluripotency. During reprogramming, a mesenchymal to epithelial transition (MET) is important for reprogramming to naïve pluripotency, consistent with the essential of role of CDH1 in maintaining pluripotency.

Multiple States of Pluripotency

While there are many kinds of stem cells, pluripotent stem cells have the broadest developmental potential since they can give rise to any mature cell type in the body. A common misconception is that all pluripotent stem cell lines are equivalent. In fact, multiple states of pluripotency exist among stem cell lines. These multiple states are organized along a pluripotency spectrum, where naïve to primed corresponds to stages of differentiation during embryo development (Fig. 2.4). In some cases, the particular state of pluripotency is reflective of the developmental origins of the stem cell line, while in other cases, the particular state of pluripotency is achieved by changing the signaling environment of the stem cell line (Weinberger et al., 2016) or their polarization, discussed further below.

Pluripotent stem cell lines have been derived from mouse embryos of at least two developmental stages. Pluripotent stem cell lines derived from the preimplantation epiblast are called ESCs, while pluripotent stem cell lines derived from the postimplantation epiblast are known as epiblast stem cells (EpiSCs) (Brons et al., 2007; Evans and Kaufman, 1981; Martin, 1981; Tesar et al., 2007). These two types of pluripotent stem cell line possess distinct properties, including growth factor requirements, gene expression, and morphology, reflective of their distinct developmental origins. In addition, ESCs are known to exist in one of two interconvertible pluripotent states, termed naïve and primed (Marks et al., 2012; Niakan et al., 2010; Wray et al., 2011).

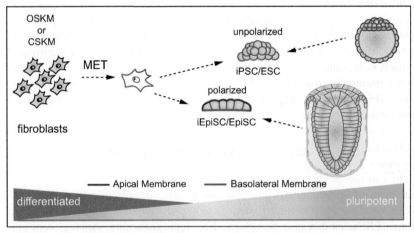

FIGURE 2.4   Cell polarization in pluripotent stem cell lines. Pluripotent stem cell lines derived from the mouse preimplantation embryo are called embryonic stem cell lines (ESCs) and are unpolarized, but adherent. ESCs can exist in a state of naïve or primed pluripotency, depending on the cell culture medium. Epiblast stem cells (EpiSCs), derived from the postimplantation embryo, are polarized and are considered more primed than are primed ESCs. In addition, either iPSCs or iEpiSCs can be derived from somatic cells, depending on the particular reprogramming conditions.

The differences between naïve and primed ESCs are numerous (Weinberger et al., 2016) and include transcriptional differences, differences in developmental potential, and X chromosome inactivation. Mouse EpiSCs and human ESCs are thought to be more primed than are primed mouse ESCs (Brons et al., 2007; Tesar et al., 2007), but both can be shifted to a more naïve state by changing the culture conditions/growth factors (Ware, 2017; Weinberger et al., 2016). Importantly, culture conditions affecting cell polarization can shift primed pluripotent stem cell lines into a more naïve state, and this is discussed further below.

Polarization of Pluripotent Stem Cell Lines

The role of cell polarization in pluripotent stem cell lines has been investigated to a limited extent. Among pluripotent stem cell lines, the degree of polarization generally correlates with the embryonic stage from which each stem cell line is derived. For example, mouse naïve and primed ESCs maintain a nonpolarized epithelial appearance, growing as dome-shaped colonies of varying sizes. Mouse ESCs lacking aPKC have been generated (Bandyopadhyay et al., 2004), indicating that cell polarity proteins are not important for maintaining their proliferative properties. In fact, interfering with *aPKC* renders mouse ESCs resistant to differentiation (Dutta et al., 2011; Mah et al., 2015), suggesting that cell polarity is necessary for exit from naïve, and entry into primed, pluripotency.

As in the preimplantation epiblast, CDH1 is localized to cell boundaries, and loss of *Cdh1* from ESCs leads to their spontaneous differentiation (Chen et al., 2010; Redmer et al., 2011; Soncin et al., 2009). Therefore, *Cdh1* helps maintain pluripotency in mouse ESCs. The mechanism by which CDH1 promotes pluripotency is an active area of inquiry, and is thought to involve stabilization of CTTNB1, since WNT signaling is important for pluripotency (Willert et al., 2003; Ying et al., 2008). In addition, one study proposed that CDH1 promotes expression and activation of pluripotency pathway components (Hawkins et al., 2012). Another study proposed that CDH1 promotes pluripotency by stabilizing OCT4 at the cell membrane, which regulates the proper dose of OCT4 (Faunes et al., 2013). These results indicate that the role of CDH1 in maintaining pluripotency is direct, rather than an indirect consequence of cell adhesion.

In contrast to mouse ESCs, mouse EpiSCs and human ESCs exist as a polarized epithelium, growing in flat colonies of varying sizes (Brons et al., 2007; Tesar et al., 2007; Thomson et al., 1998). In both mouse EpiSCs and human ESCs, inhibiting cell polarization, for example, by inhibiting ROCK and/or aPKC, promotes acquisition of properties associated with naïve pluripotency (Gafni et al., 2013; Guo et al., 2016; Ohgushi et al., 2015; Takashima et al., 2014; Theunissen and Jaenisch, 2014; Watanabe et al., 2007). Therefore, polarization of pluripotent stem cell lines is associated with a decline in pluripotency, as observed in epiblast cells during development.

While it may be the case that cell polarization is important for defining primed pluripotency, it is still unclear *why* primed pluripotent cells need to polarize, or in what way polarization would contribute to the function of pluripotent cells. In many types of adult stem cell, polarization regulates the asymmetric cell divisions that produce one differentiated and one stem cell (Moore and Jessberger, 2017). Asymmetric cell divisions of this kind are thought to be important for the functionality of adult stem cells, enabling replacement of damaged tissue without depleting the stem cell reserve (ibid.). However, pluripotent stem cell lines are not known to differentiate in this manner. Rather, pluripotent stem cell differentiation is thought to proceed in all progeny through a process of gradual changes in molecular profile, developmental potential, and proliferation, mimicking the wholesale differentiation of the epiblast. One study identified conditions that led mouse ESCs to asymmetrically express or distribute pluripotency proteins between daughter cells, suggestive of an asymmetric stem cell division (Habib et al., 2013). Unfortunately, neither the role of cell polarization nor the functional pluripotent properties of the daughter cells was scrutinized in this study. Thus, although conditions to cause asymmetric stem cell division can be engineered, the relevance of this observation as a developmental mechanism has not been established. Therefore, it remains unclear why primed pluripotent stem cells need to polarize during development, or how this prepares these cells to differentiate.

Cell Polarization During Somatic Cell Reprogramming

Pluripotent stem cell lines can also be established by reprogramming differentiated cells using a variety of genetic and chemical approaches (Takahashi and Yamanaka, 2016). For example, mouse somatic cells can be reprogrammed to iPSCs, which are very similar to mouse ESCs (Takahashi and Yamanaka, 2006). In addition, human somatic cells can be reprogrammed to human iPSCs (Takahashi et al., 2007; Yu et al., 2007). More recently, mouse somatic cells have been reprogrammed to a state similar to mouse EpiSCs (Han et al., 2011). Distinct cocktails of reprogramming factors and signaling molecules support each type of reprogramming, consistent with the roles of these factors and pathways in embryo-derived pluripotent stem cell lines (Weinberger et al., 2016).

The roles of cell polarity proteins in reprogramming has been examined to a limited extent. For example, *Cdh1* is required for reprogramming mouse somatic cells to iPSCs (Chen et al., 2010), consistent with the role of *Cdh1* in maintaining pluripotency of mouse ESCs (Bedzhov et al., 2013; Redmer et al., 2011). Curiously, however, *Cdh1* can substitute for *Oct4* as a reprogramming factor (Redmer et al., 2011). This observation is surprising because *Oct4* is a component of almost all reprogramming cocktails

(Takahashi and Yamanaka, 2016). The proposed mechanism of *Cdh1*-facilitated reprogramming is that CDH1 helps drive a MET (Lowry, 2011), known to be important for reprogramming (Li et al., 2010; Samavarchi-Tehrani et al., 2010; Shu and Pei, 2014). However, the efficiency of reprogramming was reportedly lower when *Cdh1* substituted for *Oct4* (Redmer et al., 2011), suggesting that *Oct4* provides functionality that is not completely replaced by *Cdh1*.

Knowledge that MET is involved in reprogramming mouse somatic cells to iPSCs is curious since mouse iPSCs are not overtly polarized and suggests that MET during reprogramming leads to establishment of an unpolarized epithelium. Alternatively, cells undergoing reprogramming may polarize transiently. Addressing these two possibilities at the single cell level, for example, by immunofluorescence analysis of fixed cultures, is made difficult by the fact that reprogramming efficiency is extremely low (0.1−1%), and it is not obvious at an early time point which cells would become iPSCs, given the chance. Therefore, the possibility of a transient phase of cell polarization *en route* to pluripotency remains. Recently, it was reported that fibroblasts undergo EMT before undergoing MET during an optimized reprogramming protocol (Liu et al., 2013). This observation is curious since fibroblasts are already mesenchymal and not epithelial. It should be noted that the criteria used to define epithelial and mesenchymal in this particular study were E and N-cadherin transcript levels, and not morphological changes or subcellular changes in protein localization. Further studies will be needed to identify the roles of cell polarity components in regulating pluripotent cell fate and function.

CONCLUSIONS

We have summarized literature indicating that cell polarization does not directly drive naïve pluripotency in the embryo. Similarly, cell polarization is neither essential for maintaining naïve pluripotency in stem cells nor for inducing naïve pluripotency in somatic cells. Rather, cell polarization is a hallmark of primed pluripotency. This raises the question as to what is the purpose of pluripotent cell polarization? One possibility is that cell polarization facilitates differentiation, and that the purpose of epiblast cell polarization is to prepare the epiblast for differentiation, as the epiblast undergoes EMT during gastrulation. Continued investigations into the roles of cell polarity pathway members in development will provide exciting future directions for stem cell biology.

ACKNOWLEDGMENTS

Work in our lab is supported by National Institutes of Health R01 GM104009 to AR.

REFERENCES

Alarcon, V.B., 2010. Cell polarity regulator PARD6B is essential for trophectoderm formation in the preimplantation mouse embryo. Biol. Reprod. 83, 347–358.

Anani, S., Bhat, S., Honma-Yamanaka, N., Krawchuk, D., Yamanaka, Y., 2014. Initiation of Hippo signaling is linked to polarity rather than to cell position in the pre-implantation mouse embryo. Development 141, 2813–2824.

Avilion, A.A., Nicolis, S.K., Pevny, L.H., Perez, L., Vivian, N., Lovell-Badge, R., 2003. Multipotent cell lineages in early mouse development depend on SOX2 function. Genes Dev. 17, 126–140.

Bandyopadhyay, G., Standaert, M.L., Sajan, M.P., Kanoh, Y., Miura, A., Braun, U., Kruse, F., Leitges, M., Farese, R.V., 2004. Protein kinase C-lambda knockout in embryonic stem cells and adipocytes impairs insulin-stimulated glucose transport. Mol. Endocrinol. 18, 373–383.

Barcroft, L.C., Offenberg, H., Thomsen, P., Watson, A.J., 2003. Aquaporin proteins in murine trophectoderm mediate transepithelial water movements during cavitation. Dev. Biol. 256, 342–354.

Baron, M.H., Isern, J., Fraser, S.T., 2012. The embryonic origins of erythropoiesis in mammals. Blood 119, 4828–4837.

Bedzhov, I., Zernicka-Goetz, M., 2014. Self-organizing properties of mouse pluripotent cells initiate morphogenesis upon implantation. Cell 156, 1032–1044.

Bedzhov, I., Alotaibi, H., Basilicata, M.F., Ahlborn, K., Liszewska, E., Brabletz, T., Stemmler, M.P., 2013. Adhesion, but not a specific cadherin code, is indispensable for ES cell and induced pluripotency. Stem Cell Res. 11, 1250–1263.

Boroviak, T., Loos, R., Bertone, P., Smith, A., Nichols, J., 2014. The ability of inner-cell-mass cells to self-renew as embryonic stem cells is acquired following epiblast specification. Nat. Cell Biol. 16, 516–528.

Brons, I., Smithers, L., Trotter, M., Rugg-Gunn, P., Sun, B., Chuva de Sousa Lopes, S., Howlett, S., Clarkson, A., Ahrlund-Richter, L., Pedersen, R., Vallier, L., 2007. Derivation of pluripotent epiblast stem cells from mammalian embryos. Nature 448, 191–195.

Chazaud, C., Yamanaka, Y., 2016. Lineage specification in the mouse preimplantation embryo. Development 143, 1063–1074.

Chazaud, C., Yamanaka, Y., Pawson, T., Rossant, J., 2006. Early lineage segregation between epiblast and primitive endoderm in mouse blastocysts through the Grb2-MAPK pathway. Dev. Cell 10, 615–624.

Chen, T., Yuan, D., Wei, B., Jiang, J., Kang, J., Ling, K., Gu, Y., Li, J., Xiao, L., Pei, G., 2010. E-cadherin-mediated cell-cell contact is critical for induced pluripotent stem cell generation. Stem Cells 28, 1315–1325.

Cockburn, K., Biechele, S., Garner, J., Rossant, J., 2013. The Hippo pathway member Nf2 is required for inner cell mass specification. Curr. Biol. 23, 1195–1201.

Dard, N., Le, T., Maro, B., Louvet-Vallée, S., 2009. Inactivation of aPKClambda reveals a context dependent allocation of cell lineages in preimplantation mouse embryos. PLoS One 4, e7117.

De Vries, W.N., Evsikov, A.V., Haac, B.E., Fancher, K.S., Holbrook, A.E., Kemler, R., Solter, D., Knowles, B.B., 2004. Maternal beta-catenin and E-cadherin in mouse development. Development 131, 4435–4445.

Deglincerti, A., Croft, G.F., Pietila, L.N., Zernicka-Goetz, M., Siggia, E.D., Brivanlou, A.H., 2016. Self-organization of the in vitro attached human embryo. Nature 533, 251–254.

DiZio, S.M., Tasca, R.J., 1977. Sodium-dependent amino acid transport in preimplantation mouse embryos. III. Na+-k+-atpase-linked mechanism in blastocysts. Dev. Biol. 59, 198–205.

Ducibella, T., Anderson, E., 1975. Cell shape and membrane changes in the eight-cell mouse embryo: prerequisites for morphogenesis of the blastocyst. Dev. Biol. 47, 45−58.

Ducibella, T., Albertini, D.F., Anderson, E., Biggers, J.D., 1975. The preimplantation mammalian embryo: characterization of intercellular junctions and their appearance during development. Dev. Biol. 45, 231−250.

Dutta, D., Ray, S., Home, P., Larson, M., Wolfe, M.W., Paul, S., 2011. Self-renewal versus lineage commitment of embryonic stem cells: protein kinase C signaling shifts the balance. Stem Cells 29, 618−628.

Eckert, J.J., McCallum, A., Mears, A., Rumsby, M.G., Cameron, I.T., Fleming, T.P., 2004. PKC signalling regulates tight junction membrane assembly in the pre-implantation mouse embryo. Reproduction 127, 653−667.

Eckert, J.J., Velazquez, M.A., Fleming, T.P., 2015. Cell signalling during blastocyst morphogenesis. Adv. Exp. Med. Biol. 843, 1−21.

Evans, M., Kaufman, M., 1981. Establishment in culture of pluripotential cells from mouse embryos. Nature 292, 154−156.

Faunes, F., Hayward, P., Descalzo, S.M., Chatterjee, S.S., Balayo, T., Trott, J., Christoforou, A., Ferrer-Vaquer, A., Hadjantonakis, A.K., Dasgupta, R., Arias, A.M., 2013. A membrane-associated β-catenin/Oct4 complex correlates with ground-state pluripotency in mouse embryonic stem cells. Development 140, 1171−1183.

Fierro-González, J.C., White, M.D., Silva, J.C., Plachta, N., 2013. Cadherin-dependent filopodia control preimplantation embryo compaction. Nat. Cell Biol. 15, 1424−1433.

Fleming, T.P., McConnell, J., Johnson, M.H., Stevenson, B.R., 1989. Development of tight junctions de novo in the mouse early embryo: control of assembly of the tight junction-specific protein, ZO-1. J. Cell Biol. 108, 1407−1418.

Fleming, T., 1987. A quantitative analysis of cell allocation to trophectoderm and inner cell mass in the mouse blastocyst. Dev. Biol. 119, 520−531.

Fleming, T.P., 1992. Trophectoderm Biogenesis in the Preimplantation Mouse Embryo. Springer, Netherlands.

Frum, T., Ralston, A., 2015. Cell signaling and transcription factors regulating cell fate during formation of the mouse blastocyst. Trends Genet. 31, 402−410.

Frum, T., Halbisen, M.A., Wang, C., Amiri, H., Robson, P., Ralston, A., 2013. Oct4 cell-autonomously promotes primitive endoderm development in the mouse blastocyst. Dev. Cell 25, 610−622.

Gafni, O., Weinberger, L., Mansour, A.A., Manor, Y.S., Chomsky, E., Ben-Yosef, D., Kalma, Y., Viukov, S., Maza, I., Zviran, A., Rais, Y., Shipony, Z., Mukamel, Z., Krupalnik, V., Zerbib, M., Geula, S., Caspi, I., Schneir, D., Shwartz, T., Gilad, S., Amann-Zalcenstein, D., Benjamin, S., Amit, I., Tanay, A., Massarwa, R., Novershtern, N., Hanna, J.H., 2013. Derivation of novel human ground state naive pluripotent stem cells. Nature 504, 282−286.

Gardner, R., Rossant, J., 1979. Investigation of the fate of 4-5 day post-coitum mouse inner cell mass cells by blastocyst injection. J. Embryol. Exp. Morphol. 52, 141−152.

Gardner, R., 1982. Investigation of cell lineage and differentiation in the extraembryonic endoderm of the mouse embryo. J. Embryol. Exp. Morphol. 68, 175−198.

Gardner, R., 1984. An in situ cell marker for clonal analysis of development of the extraembryonic endoderm in the mouse. J. Embryol. Exp. Morphol. 80, 251−288.

Gerbe, F., Cox, B., Rossant, J., Chazaud, C., 2008. Dynamic expression of Lrp2 pathway members reveals progressive epithelial differentiation of primitive endoderm in mouse blastocyst. Dev. Biol. 313, 594−602.

Grabarek, J.B., Zyzyńska, K., Saiz, N., Piliszek, A., Frankenberg, S., Nichols, J., Hadjantonakis, A.K., Plusa, B., 2012. Differential plasticity of epiblast and primitive endoderm precursors within the ICM of the early mouse embryo. Development 139, 129–139.

Guo, G., Huss, M., Tong, G., Wang, C., Li Sun, L., Clarke, N., Robson, P., 2010. Resolution of cell fate decisions revealed by single-cell gene expression analysis from zygote to blastocyst. Dev. Cell 18, 675–685.

Guo, G., von Meyenn, F., Santos, F., Chen, Y., Reik, W., Bertone, P., Smith, A., Nichols, J., 2016. Naive pluripotent stem cells derived directly from isolated cells of the human inner cell mass. Stem Cell Rep. 6, 437–446.

Habib, S.J., Chen, B.C., Tsai, F.C., Anastassiadis, K., Meyer, T., Betzig, E., Nusse, R., 2013. A localized Wnt signal orients asymmetric stem cell division in vitro. Science 339, 1445–1448.

Halder, G., Johnson, R.L., 2011. Hippo signaling: growth control and beyond. Development 138, 9–22.

Han, D.W., Greber, B., Wu, G., Tapia, N., Araúzo-Bravo, M.J., Ko, K., Bernemann, C., Stehling, M., Schöler, H.R., 2011. Direct reprogramming of fibroblasts into epiblast stem cells. Nat. Cell Biol. 13, 66–71.

Hawkins, K., Mohamet, L., Ritson, S., Merry, C.L., Ward, C.M., 2012. E-cadherin and, in its absence, N-cadherin promotes Nanog expression in mouse embryonic stem cells via STAT3 phosphorylation. Stem Cells 30, 1842–1851.

Hirate, Y., Hirahara, S., Inoue, K., Suzuki, A., Alarcon, V.B., Akimoto, K., Hirai, T., Hara, T., Adachi, M., Chida, K., Ohno, S., Marikawa, Y., Nakao, K., Shimono, A., Sasaki, H., 2013. Polarity-dependent distribution of angiomotin localizes Hippo signaling in preimplantation embryos. Curr. Biol. 23, 1181–1194.

Hirate, Y., Hirahara, S., Inoue, K., Kiyonari, H., Niwa, H., Sasaki, H., 2015. Par-aPKC-dependent and -independent mechanisms cooperatively control cell polarity, Hippo signaling, and cell positioning in 16-cell stage mouse embryos. Dev. Growth Differ. 57, 544–556.

Houliston, E., Pickering, S.J., Maro, B., 1987. Redistribution of microtubules and pericentriolar material during the development of polarity in mouse blastomeres. J. Cell Biol. 104, 1299–1308.

Hyafil, F., Morello, D., Babinet, C., Jacob, F., 1980. A cell surface glycoprotein involved in the compaction of embryonal carcinoma cells and cleavage stage embryos. Cell 21, 927–934.

Jedrusik, A., Parfitt, D., Guo, G., Skamagki, M., Grabarek, J., Johnson, M., Robson, P., Zernicka-Goetz, M., 2008. Role of Cdx2 and cell polarity in cell allocation and specification of trophectoderm and inner cell mass in the mouse embryo. Genes Dev. 22, 2692–2706.

Johnson, A.D., Alberio, R., 2015. Primordial germ cells: the first cell lineage or the last cells standing? Development 142, 2730–2739.

Johnson, M.H., Maro, B., 1984. The distribution of cytoplasmic actin in mouse 8-cell blastomeres. J. Embryol. Exp. Morphol. 82, 97–117.

Johnson, M.H., Ziomek, C.A., 1981a. Induction of polarity in mouse 8-cell blastomeres: specificity, geometry, and stability. J. Cell Biol. 91, 303–308.

Johnson, M.H., Ziomek, C.A., 1981b. The foundation of two distinct cell lineages within the mouse morula. Cell 24, 71–80.

Johnson, M.H., Maro, B., Takeichi, M., 1986. The role of cell adhesion in the synchronization and orientation of polarization in 8-cell mouse blastomeres. J. Embryol. Exp. Morphol. 93, 239–255.

Kang, M., Piliszek, A., Artus, J., Hadjantonakis, A.K., 2013. FGF4 is required for lineage restriction and salt-and-pepper distribution of primitive endoderm factors but not their initial expression in the mouse. Development 140, 267−279.

Kidder, G.M., McLachlin, J.R., 1985. Timing of transcription and protein synthesis underlying morphogenesis in preimplantation mouse embryos. Dev. Biol. 112, 265−275.

Kono, K., Tamashiro, D.A., Alarcon, V.B., 2014. Inhibition of RHO-ROCK signaling enhances ICM and suppresses TE characteristics through activation of Hippo signaling in the mouse blastocyst. Dev. Biol. 394, 142−155.

Krupa, M., Mazur, E., Szczepańska, K., Filimonow, K., Maleszewski, M., Suwińska, A., 2014. Allocation of inner cells to epiblast vs primitive endoderm in the mouse embryo is biased but not determined by the round of asymmetric divisions (8→16- and 16→32-cells). Dev. Biol. 385, 136−148.

Labosky, P.A., Barlow, D.P., Hogan, B.L., 1994. Mouse embryonic germ (EG) cell lines: transmission through the germline and differences in the methylation imprint of insulin-like growth factor 2 receptor (Igf2r) gene compared with embryonic stem (ES) cell lines. Development 120, 3197−3204.

Lawson, K.A., Pedersen, R.A., 1992. Clonal analysis of cell fate during gastrulation and early neurulation in the mouse. Ciba Found. Symp. 165, 3−21 discussion 21−26.

Le Bin, G.C., Muñoz-Descalzo, S., Kurowski, A., Leitch, H., Lou, X., Mansfield, W., Etienne-Dumeau, C., Grabole, N., Mulas, C., Niwa, H., Hadjantonakis, A.K., Nichols, J., 2014. Oct4 is required for lineage priming in the developing inner cell mass of the mouse blastocyst. Development 141, 1001−1010.

Leitch, H.G., Smith, A., 2013. The mammalian germline as a pluripotency cycle. Development 140, 2495−2501.

Leitch, H.G., Okamura, D., Durcova-Hills, G., Stewart, C.L., Gardner, R.L., Matsui, Y., Papaioannou, V.E., 2014. On the fate of primordial germ cells injected into early mouse embryos. Dev. Biol. 385, 155−159.

Leung, C.Y., Zernicka-Goetz, M., 2013. Angiomotin prevents pluripotent lineage differentiation in mouse embryos via Hippo pathway-dependent and -independent mechanisms. Nat. Commun. 4, 2251.

Levy, J.B., Johnson, M.H., Goodall, H., Maro, B., 1986. The timing of compaction: control of a major developmental transition in mouse early embryogenesis. J. Embryol. Exp. Morphol. 95, 213−237.

Li, R., Liang, J., Ni, S., Zhou, T., Qing, X., Li, H., He, W., Chen, J., Li, F., Zhuang, Q., Qin, B., Xu, J., Li, W., Yang, J., Gan, Y., Qin, D., Feng, S., Song, H., Yang, D., Zhang, B., Zeng, L., Lai, L., Esteban, M.A., Pei, D., 2010. A mesenchymal-to-epithelial transition initiates and is required for the nuclear reprogramming of mouse fibroblasts. Cell Stem Cell 7, 51−63.

Liu, X., Sun, H., Qi, J., Wang, L., He, S., Liu, J., Feng, C., Chen, C., Li, W., Guo, Y., Qin, D., Pan, G., Chen, J., Pei, D., Zheng, H., 2013. Sequential introduction of reprogramming factors reveals a time-sensitive requirement for individual factors and a sequential EMT-MET mechanism for optimal reprogramming. Nat. Cell Biol. 15, 829−838.

Lokken, A.A., Ralston, A., 2016. The genetic regulation of cell fate during preimplantation mouse development. Curr. Top. Dev. Biol. 120, 173−202.

Louvet, S., Aghion, J., Santa-Maria, A., Mangeat, P., Maro, B., 1996. Ezrin becomes restricted to outer cells following asymmetrical division in the preimplantation mouse embryo. Dev. Biol. 177, 568−579.

Lowry, W.E., 2011. E-cadherin, a new mixer in the Yamanaka cocktail. EMBO Rep. 12, 613−614.

Magnuson, T., Demsey, A., Stackpole, C.W., 1977. Characterization of intercellular junctions in the preimplantation mouse embryo by freeze-fracture and thin-section electron microscopy. Dev. Biol. 61, 252−261.

Mah, I.K., Soloff, R., Hedrick, S.M., Mariani, F.V., 2015. Atypical PKC-iota controls stem cell expansion via regulation of the notch pathway. Stem Cell Rep. 5, 866−880.

Maître, J.L., Niwayama, R., Turlier, H., Nédélec, F., Hiiragi, T., 2015. Pulsatile cell-autonomous contractility drives compaction in the mouse embryo. Nat. Cell Biol. 17, 849−855.

Maître, J.L., Turlier, H., Illukkumbura, R., Eismann, B., Niwayama, R., Nédélec, F., Hiiragi, T., 2016. Asymmetric division of contractile domains couples cell positioning and fate specification. Nature 536, 344−348.

Marks, H., Kalkan, T., Menafra, R., Denissov, S., Jones, K., Hofemeister, H., Nichols, J., Kranz, A., Stewart, A.F., Smith, A., Stunnenberg, H.G., 2012. The transcriptional and epigenomic foundations of ground state pluripotency. Cell 149, 590−604.

Martin, G., 1981. Isolation of a pluripotent cell line from early mouse embryos cultured in medium conditioned by teratocarcinoma stem cells. Proc. Natl. Acad. Sci. U.S.A. 78, 7634−7638.

Matsui, Y., Zsebo, K., Hogan, B.L., 1992. Derivation of pluripotential embryonic stem cells from murine primordial germ cells in culture. Cell 70, 841−847.

McDole, K., Zheng, Y., 2012. Generation and live imaging of an endogenous Cdx2 reporter mouse line. Genesis 50, 775−782.

McDole, K., Xiong, Y., Iglesias, P.A., Zheng, Y., 2011. Lineage mapping the pre-implantation mouse embryo by two-photon microscopy, new insights into the segregation of cell fates. Dev. Biol. 355, 239−249.

Meilhac, S., Adams, R., Morris, S., Danckaert, A., Le Garrec, J., Zernicka-Goetz, M., 2009. Active cell movements coupled to positional induction are involved in lineage segregation in the mouse blastocyst. Dev. Biol. 331, 210−221.

Mihajlović, A.I., Bruce, A.W., 2016. Rho-associated protein kinase regulates subcellular localisation of Angiomotin and Hippo-signalling during preimplantation mouse embryo development. Reprod. Biomed. Online 33, 381−390.

Mihajlović, A.I., Thamodaran, V., Bruce, A.W., 2015. The first two cell-fate decisions of preimplantation mouse embryo development are not functionally independent. Sci. Rep. 5, 15034.

Mitsui, K., Tokuzawa, Y., Itoh, H., Segawa, K., Murakami, M., Takahashi, K., Maruyama, M., Maeda, M., Yamanaka, S., 2003. The homeoprotein Nanog is required for maintenance of pluripotency in mouse epiblast and ES cells. Cell 113, 631−642.

Moore, D.L., Jessberger, S., 2017. Creating age asymmetry: consequences of inheriting damaged goods in mammalian cells. Trends Cell Biol. 27, 82−92.

Moore, R., Tao, W., Smith, E.R., Xu, X.X., 2014. The primitive endoderm segregates from the epiblast in β1 integrin-deficient early mouse embryos. Mol. Cell Biol. 34, 560−572.

Morris, S.A., Teo, R.T., Li, H., Robson, P., Glover, D.M., Zernicka-Goetz, M., 2010. Origin and formation of the first two distinct cell types of the inner cell mass in the mouse embryo. Proc. Natl. Acad. Sci. U.S.A. 107, 6364−6369.

Moya, I.M., Halder, G., 2016. The Hippo pathway in cellular reprogramming and regeneration of different organs. Curr. Opin. Cell Biol. 43, 62−68.

Nance, J., 2014. Getting to know your neighbor: cell polarization in early embryos. J. Cell Biol. 206, 823−832.

Niakan, K.K., Ji, H., Maehr, R., Vokes, S.A., Rodolfa, K.T., Sherwood, R.I., Yamaki, M., Dimos, J.T., Chen, A.E., Melton, D.A., McMahon, A.P., Eggan, K., 2010. Sox17 promotes differentiation in mouse embryonic stem cells by directly regulating extraembryonic gene expression and indirectly antagonizing self-renewal. Genes Dev. 24, 312−326.

Nichols, J., Zevnik, B., Anastassiadis, K., Niwa, H., Klewe-Nebenius, D., Chambers, I., Schöler, H., Smith, A., 1998. Formation of pluripotent stem cells in the mammalian embryo depends on the POU transcription factor Oct4. Cell 95, 379−391.

Nishioka, N., Yamamoto, S., Kiyonari, H., Sato, H., Sawada, A., Ota, M., Nakao, K., Sasaki, H., 2008. Tead4 is required for specification of trophectoderm in pre-implantation mouse embryos. Mech. Dev. 125, 270−283.

Nishioka, N., Inoue, K., Adachi, K., Kiyonari, H., Ota, M., Ralston, A., Yabuta, N., Hirahara, S., Stephenson, R.O., Ogonuki, N., Makita, R., Kurihara, H., Morin-Kensicki, E.M., Nojima, H., Rossant, J., Nakao, K., Niwa, H., Sasaki, H., 2009. The Hippo signaling pathway components Lats and Yap pattern Tead4 activity to distinguish mouse trophectoderm from inner cell mass. Dev. Cell 16, 398−410.

Niwa, H., Miyazaki, J., Smith, A., 2000. Quantitative expression of Oct-3/4 defines differentiation, dedifferentiation or self-renewal of ES cells. Nat. Genet. 24, 372−376.

Ohgushi, M., Minaguchi, M., Sasai, Y., 2015. Rho-signaling-directed YAP/TAZ activity underlies the long-term survival and expansion of human embryonic stem cells. Cell Stem Cell 17, 448−461.

Ohsugi, M., Hwang, S.Y., Butz, S., Knowles, B.B., Solter, D., Kemler, R., 1996. Expression and cell membrane localization of catenins during mouse preimplantation development. Dev. Dyn. 206, 391−402.

Palmieri, S., Peter, W., Hess, H., Schöler, H., 1994. Oct-4 transcription factor is differentially expressed in the mouse embryo during establishment of the first two extraembryonic cell lineages involved in implantation. Dev. Biol. 166, 259−267.

Pauken, C.M., Capco, D.G., 2000. The expression and stage-specific localization of protein kinase C isotypes during mouse preimplantation development. Dev. Biol. 223, 411−421.

Pedersen, R.A., Wu, K., Bałakier, H., 1986. Origin of the inner cell mass in mouse embryos: cell lineage analysis by microinjection. Dev. Biol. 117, 581−595.

Peyriéras, N., Hyafil, F., Louvard, D., Ploegh, H.L., Jacob, F., 1983. Uvomorulin: a nonintegral membrane protein of early mouse embryo. Proc. Natl. Acad. Sci. U.S.A. 80, 6274−6277.

Plusa, B., Frankenberg, S., Chalmers, A., Hadjantonakis, A., Moore, C., Papalopulu, N., Papaioannou, V., Glover, D., Zernicka-Goetz, M., 2005. Downregulation of Par3 and aPKC function directs cells towards the ICM in the preimplantation mouse embryo. J. Cell Sci. 118, 505−515.

Plusa, B., Piliszek, A., Frankenberg, S., Artus, J., Hadjantonakis, A., 2008. Distinct sequential cell behaviours direct primitive endoderm formation in the mouse blastocyst. Development 135, 3081−3091.

Pratt, H.P., 1985. Membrane organization in the preimplantation mouse embryo. J. Embryol. Exp. Morphol. 90, 101−121.

Ralston, A., Rossant, J., 2008. Cdx2 acts downstream of cell polarization to cell-autonomously promote trophectoderm fate in the early mouse embryo. Dev. Biol. 313, 614−629.

Ralston, A., Rossant, J., 2010. The genetics of induced pluripotency. Reproduction 139, 35−44.

Ralston, A., Cox, B., Nishioka, N., Sasaki, H., Chea, E., Rugg-Gunn, P., Guo, G., Robson, P., Draper, J., Rossant, J., 2010. Gata3 regulates trophoblast development downstream of Tead4 and in parallel to Cdx2. Development 137, 395−403.

Redmer, T., Diecke, S., Grigoryan, T., Quiroga-Negreira, A., Birchmeier, W., Besser, D., 2011. E-cadherin is crucial for embryonic stem cell pluripotency and can replace OCT4 during somatic cell reprogramming. EMBO Rep. 12, 720−726.

Renfree, M.B., 2015. Embryonic diapause and maternal recognition of pregnancy in diapausing mammals. Adv. Anat. Embryol. Cell Biol. 216, 239−252.

Resnick, J.L., Bixler, L.S., Cheng, L., Donovan, P.J., 1992. Long-term proliferation of mouse primordial germ cells in culture. Nature 359, 550–551.

Rivera-Pérez, J.A., Hadjantonakis, A.K., 2014. The dynamics of morphogenesis in the early mouse embryo. Cold Spring Harb. Perspect. Biol. 7 (11).

Saiz, N., Grabarek, J.B., Sabherwal, N., Papalopulu, N., Plusa, B., 2013. Atypical protein kinase C couples cell sorting with primitive endoderm maturation in the mouse blastocyst. Development 140, 4311–4322.

Samarage, C.R., White, M.D., Álvarez, Y.D., Fierro-González, J.C., Henon, Y., Jesudason, E.C., Bissiere, S., Fouras, A., Plachta, N., 2015. Cortical tension allocates the first inner cells of the mammalian embryo. Dev. Cell 34, 435–447.

Samavarchi-Tehrani, P., Golipour, A., David, L., Sung, H.K., Beyer, T.A., Datti, A., Woltjen, K., Nagy, A., Wrana, J.L., 2010. Functional genomics reveals a BMP-driven mesenchymal-to-epithelial transition in the initiation of somatic cell reprogramming. Cell Stem Cell 7, 64–77.

Schrode, N., Saiz, N., Di Talia, S., Hadjantonakis, A.K., 2014. GATA6 levels modulate primitive endoderm cell fate choice and timing in the mouse blastocyst. Dev. Cell 29, 454–467.

Schultz, R.M., 1993. Regulation of zygotic gene activation in the mouse. Bioessays 15, 531–538.

Shahbazi, M.N., Jedrusik, A., Vuoristo, S., Recher, G., Hupalowska, A., Bolton, V., Fogarty, N.M., Campbell, A., Devito, L.G., Ilic, D., Khalaf, Y., Niakan, K.K., Fishel, S., Zernicka-Goetz, M., 2016. Self-organization of the human embryo in the absence of maternal tissues. Nat. Cell Biol. 18, 700–708.

Shu, X., Pei, D., 2014. The function and regulation of mesenchymal-to-epithelial transition in somatic cell reprogramming. Curr. Opin. Genet. Dev. 28, 32–37.

Smyth, N., Vatansever, H.S., Murray, P., Meyer, M., Frie, C., Paulsson, M., Edgar, D., 1999. Absence of basement membranes after targeting the LAMC1 gene results in embryonic lethality due to failure of endoderm differentiation. J. Cell Biol. 144, 151–160.

Sobel, J.S., 1983a. Cell-cell contact modulation of myosin organization in the early mouse embryo. Dev. Biol. 100, 207–213.

Sobel, J.S., 1983b. Localization of myosin in the preimplantation mouse embryo. Dev. Biol. 95, 227–231.

Soncin, F., Mohamet, L., Eckardt, D., Ritson, S., Eastham, A.M., Bobola, N., Russell, A., Davies, S., Kemler, R., Merry, C.L., Ward, C.M., 2009. Abrogation of E-cadherin-mediated cell-cell contact in mouse embryonic stem cells results in reversible LIF-independent self-renewal. Stem Cells 27, 2069–2080.

Srinivas, S., 2006. The anterior visceral endoderm-turning heads. Genesis 44, 565–572.

Stephenson, R.O., Yamanaka, Y., Rossant, J., 2010. Disorganized epithelial polarity and excess trophectoderm cell fate in preimplantation embryos lacking E-cadherin. Development 137, 3383–3391.

Strumpf, D., Mao, C.A., Yamanaka, Y., Ralston, A., Chawengsaksophak, K., Beck, F., Rossant, J., 2005. Cdx2 is required for correct cell fate specification and differentiation of trophectoderm in the mouse blastocyst. Development 132, 2093–2102.

Sutherland, A.E., Speed, T.P., Calarco, P.G., 1990. Inner cell allocation in the mouse morula: the role of oriented division during fourth cleavage. Dev. Biol. 137, 13–25.

Suwińska, A., Czołowska, R., Ozdzeński, W., Tarkowski, A.K., 2008. Blastomeres of the mouse embryo lose totipotency after the fifth cleavage division: expression of Cdx2 and Oct4 and developmental potential of inner and outer blastomeres of 16- and 32-cell embryos. Dev. Biol. 322, 133–144.

Takahashi, K., Yamanaka, S., 2006. Induction of pluripotent stem cells from mouse embryonic and adult fibroblast cultures by defined factors. Cell 126, 663–676.

Takahashi, K., Yamanaka, S., 2016. A decade of transcription factor-mediated reprogramming to pluripotency. Nat. Rev. Mol. Cell Biol. 17, 183–193.

Takahashi, K., Tanabe, K., Ohnuki, M., Narita, M., Ichisaka, T., Tomoda, K., Yamanaka, S., 2007. Induction of pluripotent stem cells from adult human fibroblasts by defined factors. Cell 131, 861–872.

Takashima, Y., Guo, G., Loos, R., Nichols, J., Ficz, G., Krueger, F., Oxley, D., Santos, F., Clarke, J., Mansfield, W., Reik, W., Bertone, P., Smith, A., 2014. Resetting transcription factor control circuitry toward ground-state pluripotency in human. Cell 158, 1254–1269.

Tam, P.P., Behringer, R.R., 1997. Mouse gastrulation: the formation of a mammalian body plan. Mech. Dev. 68, 3–25.

Tao, H., Inoue, K., Kiyonari, H., Bassuk, A.G., Axelrod, J.D., Sasaki, H., Aizawa, S., Ueno, N., 2012. Nuclear localization of Prickle2 is required to establish cell polarity during early mouse embryogenesis. Dev. Biol. 364, 138–148.

Tarkowski, A.K., Wróblewska, J., 1967. Development of blastomeres of mouse eggs isolated at the 4- and 8-cell stage. J. Embryol. Exp. Morphol. 18, 155–180.

Tarkowski, A.K., Suwińska, A., Czołowska, R., Ożdżeński, W., 2010. Individual blastomeres of 16- and 32-cell mouse embryos are able to develop into foetuses and mice. Dev. Biol. 348, 190–198.

Tesar, P., Chenoweth, J., Brook, F., Davies, T., Evans, E., Mack, D., Gardner, R., McKay, R., 2007. New cell lines from mouse epiblast share defining features with human embryonic stem cells. Nature 448, 196–199.

Theunissen, T.W., Jaenisch, R., 2014. Molecular control of induced pluripotency. Cell Stem Cell 14, 720–734.

Thomson, J., Itskovitz-Eldor, J., Shapiro, S., Waknitz, M., Swiergiel, J., Marshall, V., Jones, J., 1998. Embryonic stem cell lines derived from human blastocysts. Science 282, 1145–1147.

Torres-Padilla, M., Parfitt, D., Kouzarides, T., Zernicka-Goetz, M., 2007. Histone arginine methylation regulates pluripotency in the early mouse embryo. Nature 445, 214–218.

Toyooka, Y., Oka, S., Fujimori, T., 2016. Early preimplantation cells expressing Cdx2 exhibit plasticity of specification to TE and ICM lineages through positional changes. Dev. Biol. 411, 50–60.

Tremblay, A.M., Camargo, F.D., 2012. Hippo signaling in mammalian stem cells. Semin. Cell Dev. Biol. 23, 818–826.

Vestweber, D., Gossler, A., Boller, K., Kemler, R., 1987. Expression and distribution of cell adhesion molecule uvomorulin in mouse preimplantation embryos. Dev. Biol. 124, 451–456.

Vinot, S., Le, T., Ohno, S., Pawson, T., Maro, B., Louvet-Vallée, S., 2005. Asymmetric distribution of PAR proteins in the mouse embryo begins at the 8-cell stage during compaction. Dev. Biol. 282, 307–319.

Ware, C.B., 2017. Concise review: lessons from naïve human pluripotent cells. Stem Cells 35, 35–41.

Watanabe, K., Ueno, M., Kamiya, D., Nishiyama, A., Matsumura, M., Wataya, T., Takahashi, J.B., Nishikawa, S., Muguruma, K., Sasai, Y., 2007. A ROCK inhibitor permits survival of dissociated human embryonic stem cells. Nat. Biotechnol. 25, 681–686.

Watanabe, T., Biggins, J.S., Tannan, N.B., Srinivas, S., 2014. Limited predictive value of blastomere angle of division in trophectoderm and inner cell mass specification. Development 141, 2279–2288.

Watson, A.J., Kidder, G.M., 1988. Immunofluorescence assessment of the timing of appearance and cellular distribution of Na/K-ATPase during mouse embryogenesis. Dev. Biol. 126, 80–90.

Weinberger, L., Ayyash, M., Novershtern, N., Hanna, J.H., 2016. Dynamic stem cell states: naive to primed pluripotency in rodents and humans. Nat. Rev. Mol. Cell Biol. 17, 155–169.

Wicklow, E., Blij, S., Frum, T., Hirate, Y., Lang, R.A., Sasaki, H., Ralston, A., 2014. HIPPO pathway members restrict SOX2 to the inner cell mass where it promotes ICM fates in the mouse blastocyst. PLoS Genet. 10, e1004618.

Willert, K., Brown, J.D., Danenberg, E., Duncan, A.W., Weissman, I.L., Reya, T., Yates, J.R., Nusse, R., 2003. Wnt proteins are lipid-modified and can act as stem cell growth factors. Nature 423, 448–452.

Winkel, G.K., Ferguson, J.E., Takeichi, M., Nuccitelli, R., 1990. Activation of protein kinase C triggers premature compaction in the four-cell stage mouse embryo. Dev. Biol. 138, 1–15.

Wray, J., Kalkan, T., Gomez-Lopez, S., Eckardt, D., Cook, A., Kemler, R., Smith, A., 2011. Inhibition of glycogen synthase kinase-3 alleviates Tcf3 repression of the pluripotency network and increases embryonic stem cell resistance to differentiation. Nat. Cell Biol. 13, 838–845.

Yamanaka, Y., Lanner, F., Rossant, J., 2010. FGF signal-dependent segregation of primitive endoderm and epiblast in the mouse blastocyst. Development 137, 715–724.

Yang, D.H., Smith, E.R., Roland, I.H., Sheng, Z., He, J., Martin, W.D., Hamilton, T.C., Lambeth, J.D., Xu, X.X., 2002. Disabled-2 is essential for endodermal cell positioning and structure formation during mouse embryogenesis. Dev. Biol. 251, 27–44.

Ying, Q.L., Wray, J., Nichols, J., Batlle-Morera, L., Doble, B., Woodgett, J., Cohen, P., Smith, A., 2008. The ground state of embryonic stem cell self-renewal. Nature 453, 519–523.

Yu, J., Vodyanik, M., Smuga-Otto, K., Antosiewicz-Bourget, J., Frane, J., Tian, S., Nie, J., Jonsdottir, G., Ruotti, V., Stewart, R., Slukvin, I., Thomson, J., 2007. Induced pluripotent stem cell lines derived from human somatic cells. Science 318, 1917–1920.

Chapter 3

Cell Polarity and Asymmetric Cell Division by the Wnt Morphogen

Austin T. Baldwin, Bryan T. Phillips

The University of Iowa, Iowa City, IA, United States

INTRODUCTION

Cellular diversity is perhaps the defining feature of metazoans—having a multicellular body plan allows organisms to specialize the function of its cells to better interact and respond to their environment. However, cellular diversity is not achieved trivially. In sexual organisms, life begins with the fusion of two parental gamete cells, forming the single-celled zygote. This first cell will be the progenitor of tens, hundreds, thousands, and even the trillions proposed to be present in the human body (Bianconi et al., 2013). Many of these cells will need to adopt different fates and functions to assemble an organism capable of interacting with a heterogeneous environment, but at the same time many cells will need to have the same or similar fates to effectively fill their roles. Therefore, a developing organism not only needs its cells to take on many different roles but it also needs many cells to take on the same role. To achieve these goals, organisms adopt two major strategies: specification of entire fields of cells via morphogen gradients, and binary cell fate specification (BCFS). In this chapter, we focus on BCFS through asymmetric cell division (ACD), comparing ACD to other forms of BCFS and morphogen signaling over large congruous fields of cells. As an example, we highlight a particularly well-understood mechanism of ACD that of Wnt signaling-directed ACDs during *Caenorhabditis elegans* development.

BINARY CELL FATE SPECIFICATION VIA ASYMMETRIC CELL DIVISION

BCFS is utilized in situations where an organism developmentally distinguishes one cell from its neighbors. There are two main strategies that can be used to distinguish single cells from one another. The first strategy (Fig. 3.1A),

Cell Polarity in Development and Disease. https://doi.org/10.1016/B978-0-12-802438-6.00003-6

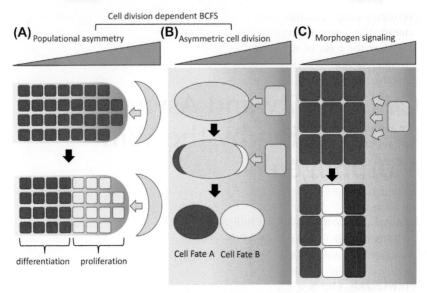

FIGURE 3.1 Morphogen, populational asymmetry, and asymmetric cell division strategies to diversify cell fate. *BCFS*, binary cell fate specification.

which could be called the "niche," occurs when a tissue uses short-range signaling to maintain all nearby cells as one type versus their neighbors, for instance, mitotic versus meiotic cells, differentiating versus dividing, or differentiating into one cell fate versus another (Lander et al., 2012). As these cells divide some daughters physically exit the range of signaling and therefore exit the niche, they differentiate from their neighbors still located in the niche (Knoblich, 2008; Neumüller and Knoblich, 2009).

Caenorhabditis elegans uses what is essentially the niche strategy to balance stem cell activity during the continuous cell divisions of its gonad. The distal tip cell (DTC) of the gonad is located at the distal tip of the gonad, and its long processes maintain contact with nearby somatic gonad cells (Hubbard, 2007). Proximity to the DTC activates the Delta/Notch signaling pathway, which inhibits entry into meiosis and promotes mitosis among germline stem cells (Hubbard, 2007). As these somatic gonad cells divide, the daughters become increasingly displaced from the DTC and eventually lose contact with it. Thus, when the daughter cells are eventually pushed from the niche by even more recently produced daughters, they escape the influence of Delta/Notch and enter meiosis, thereby "differentiating" as eggs or sperm. As this mode of BCFS relies on a prespecified cell or tissue to act as the niche, it is more commonly utilized in adult stem cell populations (Knoblich, 2008). As the individual stem cell divisions lack coordinated orientation (Crittenden et al., 2006), it seems likely that this niche system results in a "populational asymmetry" rather than polarizing individual mother cells that then divide asymmetrically. Another example of BCFS using the niche system is the mammalian intestinal crypt. Though the multipotent intestinal stem cells that reside at the base of each crypt exhibit polarized cell divisions and spindle orientation bias (Quyn et al., 2010; Bellis et al., 2012), lineage tracing experiments support the prevailing model that intestinal stem cells divide to give rise to either two new stem cells or two transit amplifying cells (Barker, 2014; Lopez-Garcia et al., 2010), suggesting that the intestinal crypt and the worm germ line both use a similar niche strategy to control symmetric self-renewing divisions within the niche that switch to differentiating divisions after the cells are crowded out of the niche.

Another method by which organisms achieve BCFS, which will be discussed here is ACD. ACD is used to distinguish the daughters of a single cell division from one another, producing two different cell types immediately after division (Knoblich, 2008; Neumüller and Knoblich, 2009) (Fig. 3.1B). ACD is utilized in two major ways: (1) to increase cellular diversity during development, and (2) to maintain populations of stem cells by maintaining stemness in one daughter of an ACD while allowing the other daughter to differentiate. The first role of ACD is implicated mainly in developmental defects, while defects in stem cell maintenance can lead to cancer progression or tissue defects (Morrison and Kimble, 2006). Thus, the importance of understanding how ACD is used to control cell fate determination is clear, as tissues require proper ACD to initially develop and then to maintain normal architecture during adulthood. ACDs are utilized extensively during development in *C. elegans* and many other, if not all, multicellular organisms. Conversely, ACD is also used to maintain homeostasis in adult organisms to renew tissues that are dependent on stem cells. Deficiencies in the ability of stem cells to self-renew are associated with aging phenotypes, and over-proliferation of stem cells is a well-known cause of cancer formation (Neumüller and Knoblich, 2009; Oh et al., 2014). Therefore, understanding the mechanisms regulating stem cell ACD using model systems like *C. elegans* is likely to be of significant clinical importance. While some stem cells utilize strategies similar to the one described above for the *C. elegans* gonad, ACD via polarization of the mother cell to produce daughters with distinct fate determinants is another widely used strategy (Knoblich, 2008).

PAR PROTEINS POLARIZE *CAENORHABDITIS ELEGANS* ZYGOTES

The polarized mother cell is more commonly used during embryonic development to generate cellular diversity. A well-known and well-researched example of this type of ACD is the first embryonic P0 cleavage of the *C. elegans* embryo. A *C. elegans* oocyte lacks preordained polarity, and the first "symmetry-breaking" event that the oocyte undergoes is fertilization (Goldstein and Hird, 1996). The sperm that fertilizes the oocyte deposits a centriole into the now-fertilized zygote, and this centriole breaks symmetry by inactivating the small GTPase RHO-1 and disrupting the previously uniform surface contractions of the oocyte (Motegi and Sugimoto, 2006; Schonegg and Hyman, 2006). This cessation of contraction then spreads across the surface of the zygote until it includes approximately 50% of the length of the zygote. In response to cessation of contractions, the various PAR proteins (partitioning defective) begin to localize asymmetrically to the cortex, with PAR-1 and PAR-2 localizing to the posterior/still cortex and PAR-3 and PAR-6 localizing to the anterior/contracting cortex (Kemphues et al., 1988; Etemad-Moghadam et al., 1995; Guo and Kemphues, 1995; Boyd et al., 1996; Watts et al., 1996; Goldstein and Macara, 2007; Nance, 2014). Polarity is then maintained by repulsion between the anterior and posterior PAR complexes, resulting in a polarized zygote. This anterior−posterior polarity then translates into asymmetries of cytoplasmic factors that result in the daughters of the zygote P0, AB and P1, adopting different cell fates. Thus, in response to the extrinsic polarizing signal of the sperm, the fertilized embryo polarizes itself and divides asymmetrically, paving the way for all the following divisions to assemble a complete animal.

DELTA/NOTCH SIGNALING DRIVES ASYMMETRIC CELL DIVISION IN FLY NEUROBLASTS

ACDs in other organisms and contexts also utilize cortical asymmetry of cell fate determinants to effect BCFS, similar to the PAR proteins. In the neuroblasts of the *Drosophila melanogaster* ventral nerve cord, the cell fate determinant *numb* localizes asymmetrically to the cortex of neuroblasts and then asymmetrically in their daughter cells (Uemura et al., 1989; Rhyu et al., 1994). *Numb* controls a transcriptional cascade via an interaction with the membrane protein *Notch*, and thus asymmetric *numb* localization is essential to asymmetric gene activation in neuroblast ACD. Asymmetric cortical localization of *numb* is controlled by a sequential series of events prior to division, which involves the "Par complex," which localizes to the apical cortex of the neuroblast and consists of the proteins Bazooka/PAR-3, PAR-6, and atypical protein kinase c (aPKC) (Kuchinke et al., 1998; Wodarz et al., 2000; Petronczki and Knoblich, 2001). Apical cortical localization of the Par complex is established during delamination of the neuroblast, which forms an apical stalk where the Par complex and other determinants of ACD localize based on contact with neighboring epithelial cells (Schaefer et al., 2000; Yu et al., 2000, 2006). aPKC phosphorylates the tumor suppressor protein lethal giant larvae (Lgl), driving localization of Lgl to the basal cortex (Ohshiro et al., 2000; Peng et al., 2000; Betschinger et al., 2003). Lgl then drives basal cortical localization of *numb* and proteins associated with *numb*, enabling *numb* to inhibit *Notch* in the basal daughter (Guo et al., 1996; Peng et al., 2000). *Notch* activates the transcription factor *tramtrack*, so asymmetric inactivation of *Notch* results in asymmetric activation of *tramtrack*, with *tramtrack* targets being activated in the apical daughter and repressed in the basal daughter (Guo et al., 1995, 1996). The apical cell retains a neuroblast fate due to *tramtrack* activation, while the basal daughter continues to divide to produce neurons or glia due to *tramtrack* inhibition (Yu et al., 2006). From this example and the previous example of PAR proteins in the *C. elegans* embryo, it can be seen that cortical asymmetry of cell fate determinants is a recurring strategy for effecting BCFS during ACD.

MORPHOGENS AS CELL POLARIZERS

One mechanism to polarize mother cells that then divide asymmetrically is through a morphogen gradient. A morphogen is defined as signaling molecules (proteins or otherwise) that act over long distances to induce responses in cells based on the concentration of morphogen that the cells interact with (Rogers and Schier, 2011). The concept of signals emanating from one tissue inducing a response in the other was developed over 100 years ago, when transplantation experiments demonstrated that relocating a piece of an embryo could influence the development of tissues surrounding the relocated piece (Lewis, 1904; Spemann and Mangold, 1923). This concept was further refined by Lewis Wolpert in his "French Flag Model," where a field of cells could respond to a graded morphogen by subdividing itself into zones dependent on the amount of morphogen each cell encountered, resulting in zones of differentiation resembling the fields of color in the French flag (Fig. 3.1C) (Wolpert, 1968; Jaeger and Martinez-Arias, 2009). Given the rather limited number of signaling molecules and signal transduction pathways compared to the vast number of developmental decisions controlled by these pathways, this concentration-dependent model has proven an attractive explanation for the induction of multiple cell fates by a single ligand (Struhl, 1989; Struhl et al., 1989; Wolpert, 1989; Tickle and Towers, 2017).

The first protein discovered that fits the mold of the theoretical morphogen was *D. melanogaster* protein bicoid. Bicoid mRNA is maternally deposited at the anterior end the *Drosophila* oocyte and eventually diffuses through the syncytial embryo, its gradient possessing the positional information to induce anterior fates at high levels of bicoid and posterior fates at lower levels of bicoid (Driever and Nusslein-Volhard, 1988a; Driever and Nüsslein-Volhard, 1988b; Struhl et al., 1989; Rogers and Schier, 2011). While bicoid patterning may not follow the strict "zoning" proposed in the French Flag model, it still provides positional information in the embryo based on the amount of bicoid in each cell. Many more examples of morphogen signaling exist, but in general all examples use morphogens to pattern and differentiate large groups of cells from other large fields of cells. However, in addition to cell fate decisions, morphogen gradients have also been demonstrated to orient target cells in their three dimensional environment (Sokol, 2015), indicating the processes of cell fate determination and cell polarity can be interwoven.

The positional information encoded in a morphogen gradient could also be used to polarize a cell in the plane of a tissue, as exemplified by the ability of a Wnt ligand, operating through the planar cell polarity (PCP) pathway, reorients a target cell's cytoskeletal architecture in the plane of the Wnt concentration gradient thus enabling coordinated cell movement during vertebrate gastrulation (Sokol, 2015) and the fly eye and wing (Carvajal-Gonzalez and Mlodzik, 2014). Wnt ligands are also the primary regulator of ACDs in *C. elegans* throughout embryonic and larval development via a signal transduction pathway that bears some similarities to Wnt/PCP but that utilizes a transcriptional output similar to the canonical Wnt/β-catenin signaling pathway (Sawa, 2012). In the well-studied Wnt/β-catenin signaling pathway target gene transcription depends on the stability of the transcriptional activator β-catenin, which is regulated in response to the Wnt morphogen (Clevers and Nusse, 2012). To regulate *C. elegans* ACD, the Wnt/β-catenin asymmetry (WβA) pathway combines β-catenin stability with cell polarity and cell division, such that the daughter cells of a Wnt-responsive mother cell will have differential response to Wnt signaling resulting in asymmetric cortical localization of Wnt signaling components and asymmetric levels of β-catenin in their nuclei (Phillips and Kimble, 2009; Sawa, 2012). As canonical Wnt/β-catenin signaling has been proposed to control ACD in some contexts in mammals (Quyn et al., 2010; Bellis et al., 2012; Habib et al., 2013), understanding how WβA regulates ACD in *C. elegans* may inform our view of Wnt-dependent ACD in other organisms. Though the specific actors vary between signaling pathways and tissues, most cellular strategies for polarization follow the logic described above—an external factor lends "directionality" to the mother cell that directionality is reinforced by internal factors, and the cell either divides or orients itself based on the asymmetry of those internal factors. The WβA pathway adapts the Wnt/β-catenin signaling pathway's ability to activate gene expression in response to an extracellular/gradient cue to asymmetric division, using that cue to polarize the cell in addition to activating appropriate target genes (Yamamoto et al., 2011).

WNT/β-CATENIN SIGNAL TRANSDUCTION

To better understand the problem of Wnt signaling in ACD in *C. elegans*, we begin with a summary of Wnt/β-catenin signaling in vertebrates and flies, where many of the gene classes that will be discussed in worms were initially identified. As mentioned previously, Wnt/β-catenin signaling revolves around the stability of β-catenin, a multifunctional protein that has roles in both adhesion and transcription. In an unsignaled cell (i.e., one that is not interacting with the extracellular Wnt ligand), cytoplasmic β-catenin levels are kept low through the action of the "destruction complex" (Stamos and Weis, 2013)

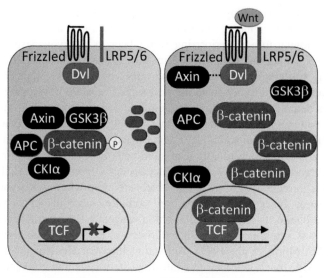

FIGURE 3.2   Model of canonical Wnt/β-catenin signaling. Left, a cell in the "unsignaled" state. Right, in the presence of the extracellular Wnt ligand, a signaling cascade commences, which results in the stabilization of β-catenin and activation of Wnt target genes. *APC*, adenomatous polyposis coli; *CK1α*, casein kinase 1α; *Dvl*, Dishevelled, GSK3β, glycogen synthase kinase 3β; *TCF*, T-cell factor.

(Fig. 3.2). The destruction complex consists of four primary proteins: the scaffolds Axin and adenomatous polyposis coli (APC) and the kinases casein kinase 1α (CK1α) and glycogen synthase kinase 3β (GSK3β) (Clevers and Nusse, 2012). Briefly, the destruction complex binds β-catenin, and β-catenin is then sequentially phosphorylated by CK1α and GSK3β. These phosphorylation events mark β-catenin for ubiquitination by the E3 ligase βTRCP, and β-catenin is then degraded by the proteasome (Clevers and Nusse, 2012).

When the transmembrane Wnt receptor Frizzled (Fz) and its coreceptor LRP5/6 are bound by the Wnt ligand, the destruction complex is inactivated. The precise mechanism for how this occurs remains controversial, but most models describe the scaffold protein Dishevelled (Dvl) and LRP5/6 as binding to and sequestering various parts of the destruction complex to either directly inhibit GSK3β or prevent the Axin interaction with β-catenin (Cselenyi et al., 2008; MacDonald et al., 2009; Taelman et al., 2010; Li et al., 2012; Kim et al., 2013). Regardless of the details of how this destruction complex inactivation occurs, the result is widely agreed on: levels of β-catenin in the cytoplasm increase. β-catenin then translocates to the nucleus, where it binds to its coactivator T-cell factor (TCF). TCF binds the promoters of Wnt target genes, and β-catenin then recruits a variety of other factors to promote expression of the Wnt target genes, which may lead to activation of cell behaviors, processes, or even fate acquisition.

Wnt/β-catenin signaling controls many processes in vertebrates. During development, Wnt signaling determines the dorsoventral axis of the embryo then transitioning to a posteriorizing factor later in embryogenesis (McCrea et al., 1993; Molenaar et al., 1996; Kim et al., 2000; Houston, 2012). Wnts have many tissue-specific effects during later vertebrate development where they collaborate with other signaling pathways to pattern developing tissues: Wnt helps define and refine the heart field, the dorsoventral axis of the neural tube and patterns the emerging limb bud among other functions (Kawakami et al., 2001; Gessert and Kühl, 2010; Hikasa and Sokol, 2013). In adults, Wnt signaling controls the proliferation of stem cells in several tissues, including the lung, breast, and colon (Tetsu and McCormick, 1999; Reguart et al., 2005) and defects in regulation of β-catenin levels are often witnessed in cancers of these tissues. Indeed, the destruction complex scaffold APC was first discovered due to its role in familial colon cancer; its involvement with Wnt signaling was subsequently revealed (Groden et al., 1991; Rubinfeld et al., 1995, 1996; Morin et al., 1997). Activating the Wnt signaling pathway by removing APC function transforms intestinal stem cells into colorectal tumor initiating stem cells (Barker et al., 2009). Knowing this, it is easy to understand the enthusiasm with which the scientific community investigated the mechanism of Wnt signaling. However, it is still unknown to what degree Wnt signaling function may be modified and adapted between tissues and cell types, even within the same organism, and how an asymmetrically presented Wnt ligand can segregate cell fate determinants and effect ACD. In one example, mammalian cells have been shown to respond to the polarized presentation of Wnt ligand by undergoing ACD where the daughters are asymmetric for Wnt pathway activation, similar to what is witnessed in WβA in *C. elegans* (Habib et al., 2013). To date, these Wnt-dependent ACDs in mammals have only been directly observed in vitro, though other data that show roles for Wnt signaling in adult mammalian neurogenesis and embryonic stem cell pluripotency suggest that Wnt/β-catenin signaling could be controlling ACD in vivo in mammals as well (Okamoto et al., 2011; ten Berge et al., 2011; Seib et al., 2013; Bielen and Houart, 2014). What then can *C. elegans* and WβA tell us about how these divisions may function within the context of a whole, live organism? Before introducing WβA, the major components of canonical Wnt/β-catenin signaling that are germane to this dissertation are described:

1. **β-catenin**—β-catenin is central to canonical Wnt signaling, which controls β-catenin stability to control transcription. In most studied organisms, β-catenin acts as both a regulator of transcription as well as an adhesive protein (Hardin, 2015); here the focus will be on its role in transcription. Initially, the β-catenin Armadillo was identified as a segment polarity gene in *D. melanogaster*, establishing β-catenin as a signaling molecule (Wieschaus and Riggleman, 1987). β-catenin was also independently

identified as a binding partner of uvomorulin/E-cadherin, highlighting its adhesive role (Ozawa et al., 1989). The discovery that β-catenin interacts with TCF in the nucleus uncovered the mechanism of β-catenin transcriptional activation (Huber et al., 1996; Molenaar et al., 1996; Schneider et al., 1996; Yost et al. 1996). β-catenin stability is regulated by a group of proteins known as the "destruction complex." In the absence of an extracellular Wnt ligand, cytoplasmic levels of β-catenin remain low as β-catenin is bound and phosphorylated by the various scaffolds and kinases of the destruction complex (see below for further details) (Stamos and Weis, 2013). Mutation of destruction complex proteins, such as APC or Axin, leads to elevation of β-catenin levels and subsequent activation of Wnt target genes and, potentially, tumorigenesis (Rubinfeld et al., 1996; Morin et al., 1997; Liu et al., 2000). Similarly, removal of the residues of β-catenin that are targeted and phosphorylated by the destruction complex leads to inappropriate stabilization of β-catenin and tumorigenesis (Morin et al., 1997). Thus, understanding the mechanisms of destruction complex targeting of β-catenin is of high importance to both developmental and cancer biology.

2. **APC**—APC is a multifunctional tumor suppressor initially identified for its role in familial colon cancer (Groden et al., 1991). Its major conserved regions include an Armadillo repeat region involved cell migration and adhesion, β-catenin and Axin-binding regions, a microtubule-binding region, and finally another region that binds to the microtubule-end-binding protein EB1 (Aoki and Taketo, 2007). These regions describe APC's two major roles as will be discussed here: β-catenin regulation and microtubule stabilization.

APC has been shown to have several different functions in β-catenin regulation. Initially, APC was found to be required for reduction of β-catenin levels in cancer cells, providing a mechanism for tumorigenesis (Munemitsu et al., 1995). It was soon discovered to act as a scaffold for Axin and GSK3β—APC binds to both cytoplasmic β-catenin and GSK3β, catalyzing β-catenin phosphorylation by GSK3β (Rubinfeld et al., 1996; Yost et al. 1996; van Noort et al., 2002; Valvezan et al., 2012). Phosphorylation of β-catenin by GSK3β then leads to its targeting for ubiquitination and degradation (Orford et al., 1997). In addition to its function in controlling β-catenin cytoplasmic levels, APC also reduces β-catenin's transcriptional activity by exporting β-catenin from the nucleus and preventing TCF binding (Neufeld et al., 2000; Rosin-Arbesfeld et al., 2003). Thus, APC is essential for maintaining low levels of nuclear β-catenin in the absence of Wnt stimulation. However, as Axin is thought to be the central scaffold of the destruction complex (Lee et al., 2003), the specific function of APC in β-catenin regulation is unclear.

APC has also been shown to regulate microtubule stability. Early on, APC was known to bind microtubules and promote their assembly

(Munemitsu et al., 1994; Smith et al., 1994). Later work showed that APC was associated with growing microtubule ends and that it bound the microtubule-associated protein EB1 (Askham et al., 2000; Mimori-Kiyosue et al., 2000). Other research went on to show that APC and EB1 were involved in microtubule organization and polymerization, as well as cell movements and migrations (Nakamura et al., 2001; Reilein and Nelson, 2005; Pfister et al., 2012). As APC mutation also results in chromosome and spindle aberrations, it seems possible that regulation of microtubules may also be essential to APC's role as a tumor suppressor (Fodde et al., 2001; Green and Kaplan, 2003). Though it has been shown that GSK3β phosphorylation of APC reduces APC's affinity for microtubules, little is known about how the β-catenin-regulating and microtubule-regulating roles of APC interact or regulated in tandem (Zumbrunn et al., 2001).

3. **Axin**—Axin is another conserved scaffold protein in the β-catenin destruction complex. Mutation of Axin leading to colon cancer is much less common than APC, but such cancers with mutant Axin have been observed (Liu et al., 2000; Webster et al., 2000; Park et al., 2005; Khan et al., 2011). Axin was initially identified as a mouse axis formation mutant and was named for the ability of Axin mRNA to cause axis inhibition in *Xenopus laevis* embryos (Zeng et al., 1997). Axin binds to both GSK3β and APC and cooperates with those proteins to reduce cytoplasmic β-catenin levels by catalyzing β-catenin phosphorylation by GSK3β (Ikeda et al., 1998; Nakamura et al., 1998). In vertebrates, it is clear that both Axin and APC bind each other and β-catenin, but as Axin catalyzes β-catenin phosphorylation, Axin is thought to be the major destruction complex scaffold (Clevers and Nusse, 2012). Importantly, Axin is thought to be essential to destruction complex function and regulation of β-catenin levels so it seems likely that the rarity of Axin mutations in cancer compared to APC is due to an essential role of Axin in various cellular and developmental contexts. Hence, perturbation of Axin activity could lead to more lethality and preclude the possibility of cancer onset. However, while Axin is thought to be the central scaffold of the destruction complex and the rate-limiting step in β-catenin degradation, the mechanism that inactivates Axin in response to Wnt signaling remains controversial, where models of Axin inactivation by membrane sequestration via phosphorylation of LRP5/6, Axin degradation via poly-ADP ribosylation and Axin intramolecular interactions regulated by phosphorylation have all been proposed (Lee et al., 2003; Cselenyi et al., 2008; Taelman et al., 2010; Li et al., 2012; Kim et al., 2013).

4. **CKIα**—The conserved function of casein kinase Iα (CKIα) in canonical Wnt/β-catenin signaling is to phosphorylate β-catenin at a single site to "prime" it for subsequent phosphorylation by GSK3β (Liu et al., 2002). The site on human β-catenin targeted by CKIα, Serine 45, has been found

to be mutated in human cancer cell lines, resulting in stabilization of β-catenin (Morin et al., 1997). While CKIα's primary negative role in Wnt/β-catenin signaling appears to be S45-phosphorylation, other CKI paralogs have been described to have a variety of both positive and negative roles, regulating the activity of many other Wnt signaling proteins (Price, 2006). Despite the varied roles of CKI kinases, it is unknown whether CKIα itself has any additional functions in Wnt signaling aside from priming β-catenin for phosphorylation by GSK3β.

5. **Dishevelled**—Dishevelled (Dvl) proteins are thought to act as the interface between the Wnt ligands and receptors at the cell membrane and the β-catenin destruction complex within the cytoplasm. Indeed, Dvl proteins sit at the center of many of the variant Wnt signaling pathways—the phrase "Wnt signaling" could accurately be replaced by "Dishevelled signaling" in most cases. In canonical Wnt/β-catenin signaling, Dvl is thought to act positively on Wnt target activation by inactivating the β-catenin destruction complex in response to Wnt ligand stimulation (Gao and Chen, 2010). Dvl proteins contain three highly conserved domains: Dsh, Egl10, Pleckstrin (DEP), PDZ, and DIX. When the Wnt ligand binds to the Fz receptor, Fz then recruits Dvl by its PDZ domain (Wong et al., 2003; Punchihewa et al., 2009). Dvl then polymerizes with itself and Axin via its DIX domain and relocates Axin to the plasma membrane, inhibiting the ability of Axin to reduce β-catenin levels in the cytoplasm (Cliffe et al., 2003; Cong et al., 2004; Schwarz-Romond et al., 2007). Despite its conservation, the function of the DEP domain in canonical Wnt/β-catenin is not as well characterized, though some data suggest that DEP also facilitates interaction with Fz (Wong et al., 2000; Tauriello et al., 2012).

The mechanism for how Dvl relocates Axin and how this in turn leads to inactivation of the β-catenin destruction complex remains controversial. One model proposes that Dvl sequesters GSK3β to endosomes to prevent complex formation and interaction with β-catenin (Taelman et al., 2010), while another proposes that the complex remains intact on Wnt activation and that β-catenin ubiquitination is merely prevented (Li et al., 2012). Thus, the role of Dvl in activating Wnt signaling remains unclear—given the function of Dvl proteins in a variety of signaling pathways, its regulation and function are likely to be highly complex and a simple model for its role in Wnt signaling may remain elusive for some time.

Many of the studies referenced above rely on in vitro experimental systems and often focus narrowly on specific interactions between Wnt signaling components. While such studies are necessary and highly useful, they often lack the context of how these interactions figure into Wnt signal transduction as a whole, and this narrow focus may fuel the controversies such as those described for β-catenin destruction complex inactivation (see the introduction to (Li et al., 2012) for a mini-review). Therefore, the lack of inclusive models of Wnt/β-catenin signaling in vivo is a weak area in our understanding of Wnt signal transduction and its potential function in ACD. A system with powerful genetics and strong in vivo imaging is ideal for this application, so we return to the nematode *C. elegans*, to determine if it can help bring any clarity to the activity of Wnt signaling in a live, intact animal.

WNT/β-CATENIN ASYMMETRY AND ASYMMETRIC CELL DIVISION IN *CAENORHABDITIS ELEGANS*

In *C. elegans*, asymmetric division of the EMS blastomere at the 4-cell stage produces both the E (Endoderm) and MS (Mesoderm) lineages in the earliest utilization of WβA signaling in worm development. The first indication that Wnt signaling was involved in this division was the discovery that the TCF homolog POP-1 was required for proper fate specification (Lin et al., 1995, 1998). Mutation of *pop-1* resulted in a conversion of the anterior MS fate to the posterior E fate, resulting in two E daughters (Lin et al., 1995). It was subsequently discovered that levels of POP-1 protein were enriched in the MS nucleus compared to that of E; thus, it was presumed that the primary function of POP-1 was repression of Wnt target genes that specify the E fate (Lin et al., 1998). A β-catenin-like protein, WRM-1, was discovered via sequence similarity, displays reciprocal asymmetry with POP-1; POP-1 levels were elevated in the MS nucleus while WRM-1 levels were reduced in MS, and vice versa in the E daughter (Meneghini et al., 1999; Rocheleau et al., 1999; Shin et al., 1999). The function of WRM-1 was found to be POP-1 nuclear export: WRM-1 promotes the interaction of POP-1 with the LIT-1 kinase and 14-3-3 protein PAR-5 to effect asymmetric nuclear export of POP-1 in E and MS (Kaletta et al., 1997; Lo et al., 2004; Yang et al., 2011). This stood in contrast to Wnt/β-catenin signaling as observed in vertebrates, flies, and even other worm tissues where TCF acts as a transcriptional repressor in the absence of Wnt signaling and β-catenin and the TCF/β-catenin interaction facilitates gene expression instead of nuclear export (Sawa and Korswagen, 2013). Due to this unique role of WRM-1/β-catenin, WβA signaling, while an interesting wrinkle on the conventional model of Wnt/β-catenin signaling, appeared to be entirely noncanonical in nature.

The genetic discovery of SYS-1, another *C. elegans* β-catenin, represented something of a return of WβA toward canonical Wnt/β-catenin signaling. Lacking significant sequence similarity, SYS-1 was identified through forward genetic screens for regulators of ACD (Miskowski et al., 2001; Siegfried et al., 2004; Kidd et al., 2005). SYS-1 has several hallmarks of a canonical β-catenin—it binds POP-1 through a conserved β-catenin-binding domain, and this interaction can activate POP-1 target genes (Kidd et al., 2005; Liu et al., 2008). In addition, despite a lack of primary sequence homology, SYS-1 and human β-catenin show a striking degree of structural conservation (Liu et al., 2008). Thus, SYS-1 appears to be the transcriptionally active β-catenin that

WRM-1 is not—mutation of *sys-1* disrupted the same ACDs as mutants of *pop-1* and *wrm-1* but *sys-1* mutation did not affect POP-1 localization, indicating that it is not involved in POP-1 export (Siegfried et al., 2004; Huang et al., 2007). SYS-1 became part of a new model of WβA signaling in which WRM-1 controlled nuclear export of POP-1 to regulate POP-1's repressive functions, while the physical interaction between POP-1 and SYS-1 allowed for target gene activation (Phillips and Kimble, 2009; Jackson and Eisenmann, 2012). However, the story of SYS-1 was far from complete, as how or whether SYS-1/β-catenin itself was regulated was unknown.

Similar to WRM-1, SYS-1 localized to nuclei of the daughters of ACDs in an asymmetric pattern, reciprocal to that of POP-1 (Fig. 3.3) (Huang et al., 2007; Phillips et al., 2007). This asymmetric pattern was controlled by conserved upstream Wnt signaling components, but how they cooperated to lower levels of SYS-1 in that anterior daughter while maintaining higher SYS-1 levels in the posterior was unclear (Huang et al., 2007; Phillips et al., 2007). From the outset, there were several lines of evidence that SYS-1 levels were likely regulated through a degradative process similar to that of canonical β-catenin: (1) SYS-1 localizes symmetrically to the centrosomes of a mother cell, indicating that each daughter initially had similar levels of SYS-1 protein (Phillips et al., 2007), (2) a *sys-1* transcriptional reporter expressing yellow fluorescent protein under the control of *sys-1* regulatory sequences is

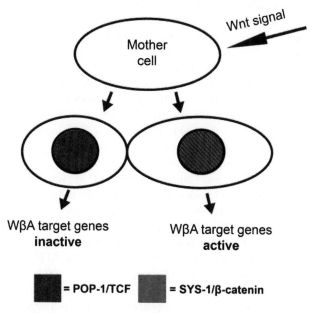

FIGURE 3.3   Basic model of Wnt/β-catenin asymmetry (WβA) signaling in *Caenorhabditis elegans*.

expressed equally in daughters of ACD, indicating regulation happens at the protein level (Phillips et al., 2007), and (3) knockdown of the proteasome via RNAi increases SYS-1 levels in the anterior daughter, indicating that SYS-1 levels are lowered through proteasomal degradation (Huang et al., 2007). Finally, this loss of proteasome phenotype was similar to the loss of APR-1, a worm APC homolog (Huang et al., 2007). Together, these lines of evidence suggested that SYS-1 regulation may depend on the function of the β-catenin destruction complex, adapted to an asymmetrically dividing cell.

Before addressing their role in SYS-1 regulation, we summarize the previously known roles of key WβA signal transducers.

1. **APR-1/APC**—APR-1 was initially identified for its sequence similarity to vertebrate APC and was so named APC-related-1. The first strong indication of APR-1's function in ACD was the discovery that it controlled the asymmetric nuclear localization of WRM-1 in the *C. elegans* asymmetrically dividing embryonic and seam cell lineages (Nakamura et al., 2005; Mizumoto and Sawa, 2007a). Further experiments demonstrated that APR-1 controls WRM-1 localization during the EMS division, via asymmetric microtubule stabilization that increases nuclear export of WRM-1 in the anterior daughter (Sugioka et al., 2011). It is notable that this microtubule-regulatory role of APR-1 appears similar to the microtubule-regulatory role of vertebrate APC described above and even involves the *C. elegans* homolog of the microtubule end-binding protein EB1 (Sugioka et al., 2011). Additionally, knockdown of APR-1 perturbed SYS-1 asymmetry in ACDs in the early embryo, indicating that it may be regulating both of the β-catenins involved in *C. elegans* ACD (Huang et al., 2007). In each studied cell division, APR-1 adopts an interesting localization pattern where it is asymmetrically presented on the anterior cortex of the mother cell and then subsequently to the cortex of the anterior daughter cell, though cytoplasmic APR-1 is still present in both daughters (Mizumoto and Sawa, 2007a; Sugioka et al., 2011; Wildwater et al., 2011). From these data, it seemed likely that APR-1 was controlling localization of both SYS-1 and WRM-1 simultaneously; however, if so, how was APR-1 dividing its function between two β-catenins?

2. **PRY-1/Axin**—PRY-1 was initially identified in screens for mutants affecting the number of sensory rays present in the tail of male *C. elegans,* where it was also identified as a negative regulator of Wnt signaling, similar to canonical Axin (Polyray-1) (Maloof et al., 1999). PRY-1 was later demonstrated to function in several Wnt signaling processes in *C. elegans*, as well as being capable of binding to APR-1 and BAR-1, another *C. elegans* β-catenin (Korswagen et al., 2002). These initial studies focused on PRY-1's role in *C. elegans* Wnt signaling separate from ACD—in these contexts, PRY-1 appears to function as a negative regulator of β-catenin-dependent processes, similar to Axin's role in canonical

Wnt/β-catenin signaling (Maloof et al., 1999; Gleason et al., 2002; Korswagen et al., 2002). The first indication that PRY-1 was involved in the WβA was the finding that *pry-1* loss of function resulted in mild defects in WRM-1 nuclear asymmetry in the EMS division (Nakamura et al., 2005). Similar to APR-1, PRY-1 was found to localize asymmetrically to the cortex of the asymmetrically dividing mother cell (Mizumoto and Sawa, 2007a). Additionally, mutation of *pry-1* resulted in disruption of APR-1 localization such that APR-1 localized symmetrically to the cortex of the dividing mother cells, placing PRY-1 upstream of APR-1 localization (Mizumoto and Sawa, 2007a). However, these same experiments show PRY-1 lacking a function in controlling WRM-1 nuclear asymmetry in early seam cell divisions (Mizumoto and Sawa, 2007a). The function of PRY-1 was then puzzling: genetically, it appeared to behave as one would predict Axin to function in canonical Wnt/β-catenin processes, but its function in WβA/ACD was unclear.

3. **KIN-19/CKIα**—KIN-19 was identified as a *C. elegans* CKIα homolog via sequence similarity with *X. laevis* CKIα (Peters et al., 1999). At the same time it was determined to regulate cell fate during the EMS, where somewhat perplexingly it was shown to positively regulate the Wnt-dependent/E cell fate, in contrast to CKIα's canonical role as a negative regulator of β-catenin (Peters et al., 1999). Little else was known about KIN-19's function in ACD in *C. elegans*, but it is worth noting a prominent piece of negative data: the *C. elegans* GSK3β homolog, GSK-3, appears to lack a function in many WβA-controlled ACDs, so it is unknown whether or not KIN-19 may be compensating for GSK3β during β-catenin regulation. Thus, whatever function KIN-19 may have in WβA signaling likely operates without GSK-3.

4. **DSH-2 and MIG-5/Dishevelled**—Though Dishevelled (Dvl) proteins are not considered to be part of the canonical β-catenin destruction complex, *C. elegans* Dvl homologs have been shown to be involved in ACD. *C. elegans* has three known Dishevelled homologs: DSH-1, DSH-2, and MIG-5. DSH-1 has been shown to be involved in mainly noncanonical Wnt signaling processes (Sanchez-Alvarez et al., 2011; Jensen et al., 2012; Huarcaya Najarro and Ackley, 2013). On the other hand, DSH-2 and MIG-5 have been shown to be involved in canonical Wnt/β-catenin signaling processes as well as Wnt-dependent ACD (Chang et al., 2005; Hawkins et al., 2005; Walston et al., 2006; Wu and Herman, 2006; Mentink et al., 2014). Specifically: DSH-2 has been shown to control asymmetric fate specification and polarity of POP-1 asymmetry in the Z1/Z4 somatic gonad precursors (Chang et al., 2005). DSH-2 has also been shown to regulate the asymmetric division of the ABpl/rppppa neuroblast, but it was later demonstrated that cell fate in this division is β-catenin independent (Hawkins et al., 2005; Hingwing et al., 2009). MIG-5 controls the asymmetric division of the male B cell and has been shown to localize

asymmetrically to the cortex of the B before and after division (Wu and Herman, 2006, 2007). Both DSH-2 and MIG-5 have been shown to be involved in cell fate specification in the asymmetric EMS division, where knockdown of *dsh-2* and *mig-5* in combination results in a reduction of cortical APR-1 on the anterior/MS daughter cortex (Bei et al., 2002; Walston et al., 2004). Interestingly, both DSH-2 and MIG-5 localize to the posterior cortex of some asymmetrically dividing mother cells in the seam cell lineage, but a loss-of-function phenotype for either *dsh-2* or *mig-5* has remained elusive in this tissue (Mizumoto and Sawa, 2007a; Banerjee et al., 2010; Sugioka et al., 2011).

While both DSH-2 and MIG-5 have been demonstrated to have roles in many ACDs in *C. elegans*, it is unclear what their mode of action is in these disparate divisions. That they have been shown to control APR-1 localization raises another question—what are the consequences of DSH-2/MIG-5 loss of function on SYS-1/WRM-1 localization? In tissues where both Dvls have been shown to control cell fate, are their functions completely redundant or is possible to parse out individualized functions for each?

One aspect of Wnt/β-catenin signaling that confounds its study in all systems is that a single gene/mutant may have subtly different functions and phenotypes in different cells/tissues, impairing the development of a cohesive model of signaling. However, the above destruction complex genes (APR-1, PRY-1, and KIN-19) have all been shown to function as regulators of cell fate specification during the WβA-dependent ACDs of a *C. elegans* epidermal tissue known as the seam (Huang et al., 2009; Banerjee et al., 2010; Gleason and Eisenmann, 2010; Ren and Zhang, 2010). Additionally, these genes have all been shown to negatively affect adoption of the Wnt-dependent seam cell fate, such that disruption of their function results in additional production of seam cells (Banerjee et al., 2010; Gleason and Eisenmann, 2010; Ren and Zhang, 2010). As these seam cell hyperplasia phenotypes are dependent on the function of SYS-1, the seam cells present the opportunity to ask whether APR-1, PRY-1, and KIN-19 are regulators of SYS-1 and better determine how these proteins had been adapted to function in ACD.

THE *CAENORHABDITIS ELEGANS* EPIDERMAL SEAM STEM CELL LINEAGE UNDERGOES REITERATIVE ASYMMETRIC CELL DIVISIONS

The *C. elegans* seam is composed of two lines of epidermal cells, one each on the left and right sides of the worm. To the observer's benefit, these are the sides presented to the observer during a standard worm slide prep (Fig. 3.4). The seam cells are essential for expression of various collagens that assemble the worm's outer cuticle (Liu et al., 1995; Thein et al., 2003). The seam cells

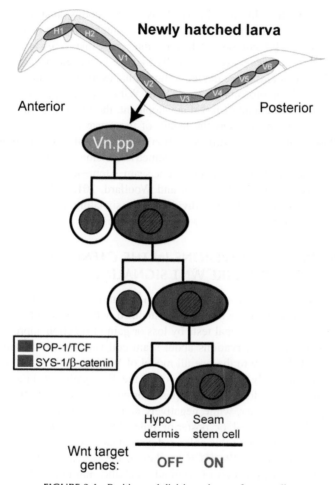

FIGURE 3.4 Position and division schema of seam cells.

also express the cuticle alae structure during L1 and adult stages (Singh and Sulston, 1978; Costa et al., 1997). Over the course of larval development, the seam cells divide several times, generally in an asymmetric pattern that produces an anterior daughter that fuses to the hypodermal syncytium hyp7, and a posterior daughter that retains the seam fate and positioning (Fig. 3.4) (Sulston and Horvitz, 1977). This "stem cell-like" mode of division allows the seam cells to continually contribute additional hypodermal nuclei to the growing worm while maintaining a consistent seam cell number and structure. While this asymmetric division is the most common mode in the seam, variants on this include a proliferative "symmetric" division during the late L1/early L2 larval stage as well other variants that produce neurons and various other cell types (Sulston and Horvitz, 1977). It is worth noting that even in the "symmetric" division where both daughters retain seam fate, all published observations of WβA component localization report that WβA appears to be active and its components asymmetrically localized, so even this "symmetric" division likely contains some degree of functional asymmetry that we have yet to ascertain (Mizumoto and Sawa, 2007a; Wildwater et al., 2011; Hughes et al., 2013). Near the end of larval development, the seam cells fuse together to form a syncytium that runs the length of the worm (Sulston and Horvitz, 1977). Despite this terminal differentiation after 2 days of larval development, some have described the reiterative asymmetric divisions of the seam cells as "stem cell-like," and the existence of a seam—stem—niche has even been proposed (Eisenmann, 2011; Brabin and Woollard, 2012). We focus on the seam cells as an asymmetrically dividing, Wnt-signaling active tissue that is amenable to both genetic and live-imaging analysis.

ASYMMETRIC CELL DIVISIONS IN THE *CAENORHABDITIS ELEGANS* SEAM REQUIRE WNT SIGNALING

One of the strengths of the seam cell system is that many members of the Wnt signal transduction pathway have been visualized with fluorescent reporters as well as subjected to functional tests by loss or gain of function approaches. In this way, many key observations about seam cell asymmetric division were made that added clarity to the overall model of WβA signaling in *C. elegans*. Reporters of positively acting members of the Wnt pathway, Frizzled, and Dishevelled cortically localize to the posterior pole of dividing mother cells in a Wnt dependent manner, while negatively acting members, Axin and APC, localize to the unsignaled or posterior pole (Mizumoto and Sawa, 2007a, 2007b). The significance of this polarized pattern has been well established for WRM-1/β-catenin and POP-1/TCF regulation: the activity of Wnt signaling disrupts posterior localization of Axin ad APC allowing only anterior localization and therefore anterior daughter cell inheritance. Thus, only the anterior daughter is thought to retain the ability to negatively regulate Wnt signaling, while the posterior daughter has decreased Axin and APC function and increased Wnt target gene expression (Gorrepati et al., 2013). In the anterior daughter, Axin stabilizes APC cortical localization and likely promotes microtubule stabilization and subsequent WRM-1 nuclear export (Sugioka et al., 2011). Low WRM-1 in the anterior nucleus in turn leads to elevated POP-1 in this nucleus and increased target gene repression (Rocheleau et al., 1999; Lo et al., 2004; Yang et al., 2011; Mila et al., 2015). Despite the appearance of asymmetric SYS-1 (high in the posterior daughter and low in the anterior daughter nuclei), the mechanism regulation of SYS-1 asymmetry, and therefore of TCF activation, was unknown.

RNAi screens investigating the function of many Wnt pathway members proved fruitful in the seam cells, as many of the cell fates controlled by this pathway have been determined by the time the dsRNA is fed to developing larvae. This allows the visualization of late seam cell divisions in an animal with reduced levels of the protein under study without perturbing earlier embryonic ACDs. These screens identified the CKIα homolog KIN-19, in addition to Axin and APC as negative regulators of seam cell fate, as their depletion increases symmetric seam cell division and terminal seam cell number (Banerjee et al., 2010; Gleason and Eisenmann, 2010). Though APC and Axin had been implicated in the negative regulation of WRM-1, responsible for lowering its nuclear levels as described above, the identification of the role of KIN-19 was intriguing because CKIα is a component of the destruction complex where it functions as a priming kinase for canonical β-catenin degradation (Liu et al., 2002). Indeed, analysis of SYS-1/β-catenin levels after loss of either KIN-19/CKIα or APR-1/APC indicates these destruction complex members negatively regulate SYS-1 in the unsignaled daughter (Baldwin and Phillips, 2014). This conserved function of destruction complex members in negatively regulating the transcriptional activator of Wnt signaling suggests functional conservation of function between nematodes, flies, and vertebrates.

Loss of PRY-1/Axin, however, results in a distinct phenotype. *pry-1* homozygotes display randomized asymmetry after seam cell division where the two daughters of an ACD typically assume either seam or hypodermal fates, but the positions of these cells can either be anterior or posteriorly located (Baldwin and Phillips, 2014; Lam and Phillips, 2017). The regulation of the direction of polarity rather than asymmetry per se suggests PRY-1/Axin helps establish mother cell polarity rather than being essential for enacting the polarity cue. The finding that this randomized polarity is rendered symmetric *pry-1*; *apr-1* double loss of function animals indicates that downstream polarity effectors, such as APC, are still functional in *pry-1* mutants. This polarity function of Axin was explained by examination of cortical APC asymmetry in *pry-1* mutants where a randomized cortical APC localization pattern inversely correlated with the randomized SYS-1 asymmetry in these mutants. Therefore, PRY-1/Axin is dispensable for destruction complex activity but functions primarily to localize APR-1 to the anterior pole of the dividing mother cell, while APR-1 negatively regulates SYS-1 and WRM-1 to promote the unsignaled fate.

Dual fluorescence labeling of SYS-1 and APR-1 in wild type and *pry-1* mutants shows that these proteins are typically mutually exclusive; low SYS-1 due corresponds to high cortical APR-1 (and vice versa), consistent with an APR-1/APC role in SYS-1 negative regulation (Baldwin and Phillips, 2014). However, this pattern of can be uncoupled by mutating *wrm-1*, which causes lowered cortical APC in both daughters yet these *wrm-1* mutants continue to show the wild-type pattern of SYS-1 asymmetry. This *wrm-1* phenotype demonstrates that the anterior daughter can lower SYS-1 even though its cortical APC has been reduced (though not eliminated). Additionally, it was unexpectedly found that KIN-19/CKIα appears to have a second role in promoting the unsignaled fate since KIN-19 depletion results in symmetrically high cortical APC. This increase in posterior APC had no effect on SYS-1, which is symmetrically high after KIN-19 depletion. Additionally, WRM-1 asymmetry does not follow the same rules as SYS-1, though its regulation has been shown to be under the control of some of the same destruction complex members (Mizumoto and Sawa, 2007b). While WRM-1 is negatively regulated by APC (Sugioka et al., 2011), WRM-1 levels symmetrically decrease in the absence of KIN-19, indicating that CKIα is a positive regulator of nuclear WRM-1. Taken together, these data are consistent with a model whereby the nuclear levels of WRM-1 and SYS-1 are regulated by distinct pools of APC (Baldwin and Phillips, 2014). Entry into the SYS-1-regulating pool requires KIN-19 and entry into the WRM-1 regulating pool require cortical WRM-1. Both pools are asymmetrically localized by Axin. The relative amount of APC in either pool appears to be regulated by upstream Wnt signaling since the double mutants of two Dishevelled paralogs, DSH-2 and MIG-5, show symmetrically decreased SYS-1, yet symmetrically increased WRM-1 (Baldwin et al., 2016). This suggests that Dvl helps partition the APC pools, consistent with its well-established role as a "gatekeeper" that dictates downstream events in varied Wnt signaling contexts (Sokol, 1999; Hingwing et al., 2009; King et al., 2009; Wallingford and Mitchell, 2011).

A working model that summarizes the above *C. elegans* ACD data are shown in Fig. 3.5. Wnt ligand is presented to a dividing mother cell in a posterior-to-anterior gradient, localizing Frizzled receptors and Dishevelled to the posterior cortex. This asymmetric pattern limits PRY-1/Axin to the anterior pole, where it recruits APR-1/APC. APC is subdivided into a WRM-1-regulating (blue) pool that stabilizes microtubules and increases WRM-1 nuclear export and a SYS-1-regulating (red) pool that degrades SYS-1. KIN-19/CKIα is required for, and may also function in, the SYS-1 regulating pool. Dishevelled controls the balance of the two pools by increasing the WRM-1-regulating pool at the expense of the SYS-1-regulating pool. Thus, after division both WRM-1 and SYS-1 are low in the anterior nucleus, while TCF is high and represses target

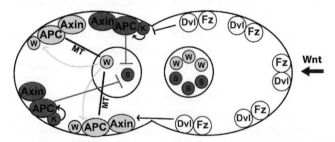

FIGURE 3.5 **Working model for Wnt/β-catenin asymmetry pathway during *Caenorhabditis elegans* asymmetric cell division.** See text for details. Anterior is to the left. *APC*, adenomatous polyposis coli.

gene expression. The posterior nucleus, lacking either β-catenin negative regulatory mechanism, accumulates both SYS-1 and WRM-1. TCF is lowered by WRM-1- and LIT-1-dependent nuclear export allowing for target gene expression.

SYS-1/β-CATENIN INHERITANCE IS LIMITED BY CENTROSOMAL DEGRADATION

Given the fact that the downstream effectors of Wnt signaling, elevated nuclear WRM-1 and SYS-1 and decreased nuclear POP-1 have important roles in cell fate determination in the daughter cells of an ACD, it seems likely that their inheritance or activity may be under tight control after the daughter cells themselves divide in the subsequent ACD, giving rise to four distinct granddaughter fates. This seems especially important in early embryonic development where ACDs can be separated by just 20 min (Sulston et al., 1983). The inheritance of one of the Wnt pathway effectors, SYS-1, has been elucidated (Vora and Phillips, 2015). SYS-1 levels are under tight control presumably because the inheritance of high levels of SYS-1 from a signaled mother would direct both daughters toward a symmetric signaled fate, representing a loss of ACD due to the specific loss of the unsignaled daughter. Analysis of fluorescently labeled SYS-1 shows that SYS-1 symmetrically localizes to mitotic centrosomes in all observed dividing cells (Phillips et al., 2007; Baldwin and Phillips, 2014; Vora and Phillips, 2015). Photobleaching experiments indicate that this centrosomal SYS-1 is rapidly turned over in a proteasome-dependent fashion. Preventing centrosomal localization leads to elevated, though still asymmetric, SYS-1 levels in both daughter cells and an increased likelihood of both daughters assuming the signaled fate (Vora and Phillips, 2015). This leads to a model whereby SYS-1 is cleared from the mother cell specifically during mitosis by centrosomal localization and subsequent proteasomal localization (Vora and Phillips, 2016). However, it is unknown what percentage of SYS-1 is turned over and whether similar clearance mechanisms exist for the other Wnt effectors. Thus, it remains possible that daughter cells still "remember" whether they are descended from a signaled or unsignaled lineage. Indeed, the discovery of SYS-1 enrichment in both daughters of a signaled cell compared to their cousins (daughters of the unsignaled cell) suggests some aspect of active Wnt signaling is retained over multiple generations (Zacharias et al., 2015; Zacharias and Murray, 2016).

The study of Wnt signaling's role in regulating ACD has described novel functions for Wnt signaling components that have not yet been described to those components in canonical Wnt/β-catenin signaling. However, ACD studies have also identified conserved functions of conserved Wnt/β-catenin signaling components having been adapted to ACD, indicating that WβA signaling can work quite similarly to canonical signaling at times. Some remaining questions raised in the next section include the following: Are Wnt signaling components too far removed from vertebrate Wnt/β-catenin signaling to have any relevance to the greater whole of Wnt signaling? Or are data here that may prove to be informative to cancer biologists and those studying vertebrate ACD? In this next section we discuss the possible species-specific "*C. elegans*/ACD specific phenomena" and also how *C. elegans* ACDs can inform the broader Wnt signaling field in "The Future of Wnt/β-catenin and ACD."

CAENORHABDITIS ELEGANS/ASYMMETRIC CELL DIVISION-SPECIFIC PHENOMENA

PRY-1/Axin Is Upstream of the Destruction Complex

The foremost difference between canonical Wnt/β-signaling and WβA signaling is that loss of PRY-1/Axin does not inhibit SYS-1/β-catenin degradation. In canonical Wnt/β-catenin signaling, Axin is thought to be the primary scaffold of the destruction complex, being responsible for both binding β-catenin and recruiting GSK3β as well as being the target of destruction complex inhibition by Dishevelled (Ikeda et al., 1998; Nakamura et al., 1998; Itoh et al., 2000; Cliffe et al., 2003). When addressing PRY-1 function, it is important to start with more discussion of the genetics involved in the above worm ACD experiments. The *pry-1* allele utilized in these experiments, *(mu38)*, is a point mutation that results in a premature stop codon and is presumed to be a null allele (Korswagen et al., 2002). This presumption is based on the hypothesis that nonsense-mediated mRNA decay (NMD) in *C. elegans* is highly efficient and would likely degrade the truncated *pry-1* transcript. In the absence of NMD, this truncated transcript would result in some partial function since it encodes the conserved Axin RGS domain, required for the APC interaction, but lacks the DIX domain, which is required for Axin's interaction with Dishevelled (Dvl) and therefore Wnt signal transduction (Itoh et al., 2000; Spink et al., 2000; Schwarz-Romond et al., 2007; Fiedler et al., 2011).

Previous work examining canonical Wnt/β-catenin signaling in *C. elegans* showed that *pry-1(mu38)* resulted in defects in migration of the QR neuroblast (Korswagen et al., 2002). QR migration is dependent on another β-catenin, BAR-1, and BAR-1-dependent processes are thought to be more similar to canonical Wnt signaling than SYS-1/WβA-dependent processes (Eisenmann et al., 1998; Maloof et al., 1999; Natarajan et al., 2001; Gleason et al., 2002; Korswagen, 2002; Korswagen et al., 2002). That *pry-1(mu38)* results in overactivation of BAR-1-dependent processes but does not negatively regulate SYS-1 levels is definitely curious, and indicates that PRY-1 may have very different functions between canonical Wnt/β-catenin signaling and WβA signaling, even within the worm. Discovering the mechanism that switches PRY-1 between SYS-1- and BAR-1-dependent processes would explain much about how Wnt/β-catenin signaling has been specialized for ACD in WβA. We suggest that posttranslational modification of PRY-1 is mediated by interactions with either Wnt receptors or Dishevelled as it has been shown that Axin activity is heavily dependent on its posttranslational modification status in vertebrates (Song et al., 2014). While it is conceivable that *pry-1(mu38)* may possess some residual activity in WβA signaling, it seems likely that PRY-1 regulation of SYS-1 is more noncanonical than its regulation of BAR-1.

Also of interest with PRY-1 is the revelation of its localization pattern in the seam cells. Given that *pry-1*'s genetic function appears to control the polarity of WβA signaling that it localizes symmetrically to the L4 seam cell cortex is quite fascinating. Even if the symmetric localization of transgenic PRY-1 is an artifact of PRY-1 overexpression, this overexpression does not result in cell fate changes, indicating that the signaling system can tolerate cortical PRY-1 symmetry. We hypothesize that PRY-1 interacts with APR-1/APC at the anterior cortex and DSH-2 and MIG-5/Dishevelled at the posterior cortex, leading to the asymmetric PRY-1 posttranslational modification mentioned above. If DSH-2/MIG-5 is anchored to the posterior cortex by an interaction with Frizzled or other Wnt receptors, an interaction between PRY-1 and DSH-2/MIG-5 that prevents a PRY-1/APR-1 interaction could effectively limit APR-1 cortical localization to the anterior daughter even if PRY-1 localizes symmetrically. Under this hypothesis, residual *pry-1(mu38)* function is an important question—the truncated PRY-1 protein would likely be able to interact with APR-1 but not Dishevelled, and the presence of depolarized SYS-1/β-catenin negative regulation in *pry-1(mu38)* is no longer a surprising result but a prediction of the model. However, how then would *pry-1(mu38)* result in Wnt target gene activation in canonical Wnt/β-catenin processes, where it could even be predicted to be unresponsive to Wnt/Dvl if it could interact with APR-1 but not Dvl (Korswagen et al., 2002)? So, once again, there is a definite difference in function between PRY-1 in WβA and canonical Wnt/β-catenin signaling in *C. elegans* and adaptation to ACD likely hinges on this difference.

Based on the difference in PRY-1 function between QR migration and seam cell specification, we suggest that the function of PRY-1 is distinct between canonical and WβA signaling. In canonical processes such as QR migration, PRY-1 functions like a canonical Axin that is directly involved in negative regulation of β-catenin levels. In WβA/ACD, PRY-1 acts as an adapter between the anterior/negative Wnt regulators (APR-1) and the posterior/positive Wnt regulators (Wnt, Wnt receptors, Dvl). How PRY-1 switches between these functions in various tissues then remains an open question. One idea is that the function of KIN-19/CKIα, GSK-3/GSK3β, DSH-2, MIG-5, or all four are responsible for switching PRY-1 between canonical functions and ACD functions. KIN-19 appears to be specific to WβA/ACD-involved processes, while GSK-3 appears to be specific to canonical Wnt/β-catenin signaling processes. One caveat with this idea is that knockdown of KIN-19 does not appear to result in an increase of canonical Wnt signaling activity, i.e., SYS-1/ β-catenin degradation. However, that result could be confounded by any direct regulation of SYS-1 that KIN-19 may affect, which we discuss next.

KIN-19 Versus GSK-3 Function in β-Catenin Regulation

To date, no function has been described for the *C. elegans* homolog of GSK3β, GSK-3, in the Wnt asymmetric division pathway. GSK-3 does function in Wnt signaling in *C. elegans*, having function demonstrated in the EMS division as well as canonical Wnt/β-catenin signaling processes (Schlesinger et al., 1999; Maduro et al., 2001; Korswagen et al., 2002; Walston et al., 2004). However, there exist no data to suggest that GSK-3 functions in asymmetric division outside of EMS, where GSK-3 may function in one of the pathways parallel to WβA that converge on cell fate in EMS. In numerous RNAi screens, we assayed for changes in the Wnt signaled fate (e.g., DTC or seam cells) as a result of *gsk-3(RNAi)* but failed to find any gains in DTCs, suggesting that GSK-3 is not a negative regulator of DTC specification or the WβA signaling pathway that is required for DTC specification (Chesney et al., 2009). If GSK-3 is essential to canonical Wnt/β-catenin signaling but dispensable to WβA signaling, what has replaced it?

We propose that the GSK-3 function in beta-catenin negative regulation appears to have been replaced by KIN-19/CKIα. Knockdown of KIN-19 results in significantly increased SYS-1/β-catenin levels in anterior seam cell daughter nuclei (Baldwin and Phillips, 2014). In flies and vertebrates, GSK3β is necessary for phosphorylating β-catenin, marking β-catenin for ubiquitination by the ubiquitin ligase βTRCP, while CKIα is responsible only for "priming" β-catenin for GSK3β phosphorylation (Hart et al., 1998; Kitagawa et al., 1999; Liu et al., 1999, 2002). If both of these steps are necessary in canonical Wnt/β-catenin signaling, why is only the CKIα step important to WβA signaling? Two hypotheses are consistent with this idea:

The first is that the "priming" phosphorylation event by KIN-19/CKIα could mark SYS-1/β-catenin for ubiquitination without requiring further phosphorylation by GSK3β. If βTRCP binds and ubiquitinates β-catenin by recognizing only the CKIα phosphorylation site, then GSK3β would no longer be necessary for negative regulation of β-catenin levels. It appears that this could be the case for KIN-19 and SYS-1. Unfortunately, additional biochemical experiments are required before it can be definitively claimed that directly KIN-19/CKIα phosphorylates SYS-1/β-catenin. Further, a ubiquitin ligase that controls WβA signaling has not yet been identified, so it is also speculative to say that SYS-1 is ubiquitinated.

A second hypothesis is that there is another cryptic kinase that phosphorylates SYS-1 to prepare it for ubiquitination. From a mechanistic standpoint, this hypothesis is not as parsimonious as the first. However, such a cryptic kinase has potential to be more relevant to canonical Wnt/β-catenin signaling if it has some function in Wnt signaling in other systems. Forward and reverse genetic screens for determinants of seam cell fate/WβA signaling do not appear to be saturated, so it seems conceivable that such a kinase may yet be identified.

THE FUTURE OF WNT/β-CATENIN AND ASYMMETRIC CELL DIVISION

APR-1 as the Key Regulator of Differential WβA Outputs

While PRY-1 is dispensable for negative regulation of nuclear β-catenin levels in the WβA pathway, APR-1/APC is most certainly required. In canonical Wnt/β-catenin signaling, the role of APC is something of a mystery: it is obviously necessary, based on cancer phenotypes in APC mutant patients and mammalian model systems, but why? Perhaps some of the confusion comes from the multifunctional nature of APC. While APC can interact with β-catenin, Axin appears to be the primary scaffold, as Axin interacts with Dishevelled, GSK-3β, and β-catenin, acting as the centerpiece of both negative and positive regulation of Wnt target genes. Various experiments have tried to place APC in greater context. For one example, APC has been shown to enhance the activity of GSK-3β for β-catenin phosphorylation (Valvezan et al., 2012). APC has also been proposed to have nuclear functions, where it may help modulate nuclear levels of β-catenin in addition to helping control β-catenin stability (Henderson, 2000; Neufeld et al., 2000; Rosin-Arbesfeld et al., 2000).

In *C. elegans* and the WβA pathway, APR-1/APC appears to have a more essential role in β-catenin regulation than in canonical signaling. While in canonical Wnt signaling APC appears to be an enhancer/modulator of Axin function, in WβA PRY-1/Axin appears to be a modulator/polarizer of APR-1/APC. APR-1 is the asymmetrically localized destruction complex protein, and APR-1 is the negative regulator of SYS-1/β-catenin levels, while PRY-1 appears to eschew either of those functions, at least in L4 seam divisions. While this may be a *C. elegans*/ACD-specific phenomenon, it is an interesting question whether or not APC has this same ability as APR-1 in vertebrate Wnt/β-catenin signaling. It is possible that vertebrate APC and APR-1 are just different in this regard, but it is equally possible that the above uncertainty about APC's primary function, compared to the more well-established roles of Axin, has masked its importance in canonical Wnt signaling. On the other hand, differences in function between APC and APR-1 could be due to differences in how they are regulated by Wnt signaling, which is one area where APR-1 has particular promise in informing our view of Wnt/β-catenin signaling.

The most interesting feature of APR-1 in the seam cells is the presence of the differentially regulated functional "pools" of APR-1 that are present in

WβA-regulated cell divisions. That these pools could be distinguished from each other through a few genetic experiments indicates that if a similar mechanism exists in vertebrates that it may be relatively simple to parse out. Therefore, investigating how APR-1 switches between SYS-1 and WRM-1 regulation could be very informative, even at the level of cancer biology. The SYS-1-regulating and WRM-1-regulating APR-1 roles very closely resemble functions of APC described in vertebrates: the SYS-1-regulating role obviously resembles β-catenin negative regulation, while WRM-1-regulation via microtubules resembles the microtubule-regulating roles of APC (Munemitsu et al., 1994; Smith et al., 1994; Askham et al., 2000; Mimori-Kiyosue et al., 2000; Nakamura et al., 2001; Zumbrunn et al., 2001; Green and Kaplan, 2003; Green et al., 2005; Reilein and Nelson, 2005; Sugioka et al., 2011; Baldwin and Phillips, 2014). However, few experiments in vertebrates or their cells have analyzed these two functions of APC in tandem; one study has shown that GSK3β phosphorylation of APC decreases APC's affinity for microtubules, and another shows that GSK3β phosphorylation of APC increases APC's affinity for β-catenin, but it has not been explored in detail how that event may be regulated by Wnt signaling or how phosphorylated and nonphosphorylated APC may function in the same cell at the same time (Rubinfeld et al., 1996; Zumbrunn et al., 2001). In WβA signaling, the opportunity exists to ask how APC can regulate β-catenin stability and microtubule stability at the same time, and how Wnt signaling controls that balance. Given that both of these functions of APC have been shown to be disrupted in APC mutant cancers, better understanding how APC moves between these functions could be of great interest to developmental and cancer biology as a whole.

Why might two functionally distinct pools of APR-1 be necessary for WβA signaling? The answer may hinge on the differences between SYS-1 and WRM-1 regulation. First, SYS-1 is controlled by degradation while WRM-1 is regulated by nuclear export, but it is still unclear precisely how APR-1 facilitates either type of regulation (Nakamura et al., 2005; Takeshita and Sawa, 2005; Huang et al., 2007; Mizumoto and Sawa, 2007a; Phillips et al., 2007; Sugioka et al., 2011). If APR-1 interacts with SYS-1, its ability to regulate WRM-1 may be inhibited and vice versa. Splitting APR-1 into SYS-1- and WRM-1-regulatory pools prior to division would prevent these functions from competing with one another. Second, the localization patterns of SYS-1 and WRM-1 are distinct at every stage of seam cell division. Again, explicitly distinguishing two pools of APR-1 would allow for APR-1 to fulfill both of its functions without having to encounter competition between centrosomal/ cytoplasmic or cortical localization. It is also unclear whether or not APR-1 has a "default pool" in the absence of Wnt regulation. Loss of KIN-19 or MIG-5 apparently shifts APR-1 from the SYS-1-regulatory pool to the WRM-1-regulatory pool, suggesting that WRM-1 regulation may be the default. Conversely, loss of both DSH-2 and MIG-5 appears to shift APR-1 toward primarily regulating SYS-1, suggesting that the SYS-1-regulatory pool is the default. The interaction of Wnt signaling components to separate the functions of APR-1 is likely to be highly complex, but based on the above data we suggest that DSH-2 and MIG-5 catalyze phosphorylation of APR-1 via KIN-19 to alter its function. In vitro studies of mammalian APC have shown that casein kinase and GSK3β phosphorylation of APC greatly increase its affinity for β-catenin (Rubinfeld et al., 2001; Ha et al., 2004; Xing et al., 2004). KIN-19 phosphorylation of APR-1 could then be hypothesized to increase APR-1's affinity for SYS-1, and this would be consistent with KIN-19 loss resulting in increased negative regulation of WRM-1. Future experiments examining the physical interactions of APR-1, KIN-19, DSH-2, and MIG-5 will likely be essential for determining how the pools of APR-1 are specified.

Dishevelled Balances Wnt/β-Catenin Asymmetry Outputs

Dishevelled is an interesting case and we believe it holds promise for providing insights into canonical Wnt/β-catenin signaling. Though DSH-2 and MIG-5 are divergent Dishevelleds, they still have a high degree of structural conservation with Dishevelleds in other clades (Dillman et al., 2013). C. elegans is far from the only organism that uses multiple Dishevelled paralogs to achieve Wnt signal transduction. Though one Dvl in mice appears to bear most of the canonical Wnt/β-catenin signaling workload, other Dvls present at lower levels function redundantly in Wnt signaling as well (Etheridge et al., 2008). It may be that murine Dvls are acting redundantly and interchangeably but equally plausible is the idea that they have distinct roles regulating specific downstream signaling events. DSH-2 and MIG-5 appear to function redundantly at the gross anatomical level, such as in seam cell specification, but their individual functions in regulating those processes may be specialized and these specialized roles include the regulation of the other Dvl paralog. In the case of SYS-1 regulation, DSH-2 and MIG-5 appear to have some effect on each other's activity, such that inhibition of SYS-1-regulating APR-1 increases in the absence of either DSH-2 or MIG-5. Are there ways to model the functions of DSH-2 and MIG-5 that align with described functions of Dvl in canonical systems?

An interesting possibility is one where MIG-5 and DSH-2 form polymers, and that different regulatory roles are accomplished by DSH-2/MIG-5 complexes that contain different amounts of either Dvl. Dvl has been shown to form polymers in response to Wnt signaling, but it should be noted that the study in question examined polymers composed of only one vertebrate Dvl, Dvl2 (Schwarz-Romond et al., 2007). A multi-Dvl complex may be necessary for DSH-2/MIG-5 to integrate both polarity cues from upstream signaling components as well as the downstream signaling cues that Dvl is expected to accomplish. In vertebrates, such a "heteropolymer" mechanism could allow Wnt signaling and Dvl to target multiple substrates. Dvls have been reported to act redundantly in mammals, but the detailed mechanisms of how this redundancy functions have not been investigated (Etheridge et al., 2008). Using the previous Dvl2 polymerization experiments as a model, the presence of DSH-2/MIG-5 polymers should be investigated.

An alternative hypothesis is that DSH-2 and MIG-5 act individually without forming complexes with each other, but that they integrate signaling information at the destruction complex independently of each other. MIG-5 could bind the destruction complex independently of DSH-2, polarizing and regulating the activity of the destruction complex, while DSH-2 separately modulates destruction complex activity. There is evidence for and against this hypothesis. For instance, overexpression of MIG-5 and DSH-2 shows that individually either one can stabilize SYS-1, and that either Dvl can form puncta at high levels of expression (Baldwin et al., 2016), showing that they can inactivate the destruction complex individually. Conversely, the activity of this overexpressed Dvl did not appear to be polarized by Wnt signaling, which is essential for proper WβA pathway function. That DSH-2 and MIG-5 seem to modulate each other's activity would suggest a physical interaction between the two of them, but this is not strictly necessary for them to have a genetic interaction. Based on the evidence that currently exists, the "heteropolymer" hypothesis remains appealing.

The appearance of the overexpressed MIG-5 and DSH-2 puncta and their ability to colocalize with and stabilize SYS-1 is also interesting when comparing WβA and the canonical pathway. It has long been part of canonical Wnt/β-catenin signaling models that signal transduction requires inactivation and/or sequestration of the destruction complex, and that this may involve formation of "signalosomes" (Bilić et al., 2007; Taelman et al., 2010; Fiedler et al., 2011; Mendoza-Topaz et al., 2011; Li et al., 2012). The MIG-5 and DSH-2 puncta are then interesting for several reasons. First, Dvl is not thought to bind β-catenin directly; therefore, if SYS-1 is colocalizing with these Dvl puncta, then some other β-catenin-binding protein is likely being recruited to these puncta. It is possible that SYS-1 is being recruited to these Dvl puncta by APR-1/APC, but PRY-1/Axin is another likely candidate for SYS-1 recruitment. Understanding what destruction complex component is targeted by DSH-2/MIG-5 puncta can inform our models of DSH-2/MIG-5 regulation of SYS-1 as well as our models of canonical signal transduction. Second, these MIG-5/DSH-2 puncta appear to function similarly to Dvl in the canonical pathway (unpolarized β-catenin stabilization) instead of how they should function in WβA (polarized β-catenin stabilization). Finally, signalosomes are typically thought to require LRP5/6, which have yet to be discovered in *C. elegans*, though the possibility of a cryptic LRP5/6 has not been exhausted. Elucidating a LRP5/6-independent signalosome formation mechanism would address the minimum requirements a cell needs to respond to a Wnt ligand. Indeed, studying what does and does not allow for formation of these puncta may allow us to answer questions about the relative roles of DSH-2 and

MIG-5, how upstream signaling is transduced through them, and how DSH-2 and MIG-5 interact to regulate WβA. For example, does overexpression of both DSH-2 and MIG-5, simultaneously at similar levels, still result in the formation? If they do, are they polarized? Do they still stabilize SYS-1? Is upstream Wnt signaling necessary for formation of the puncta? Answering these questions could provide information about how WβA transduces both β-catenin regulation and polarity signals, and perhaps even inform the "heteropolymer" hypothesis mentioned above.

While these large Dvl puncta are not likely to be that relevant to physiological levels of WβA signaling, they have the potential to answer questions about WβA signal transduction that have thus far evaded us. Overexpression experiments have many drawbacks, but MIG-5 overexpression would allow for modeling of Wnt/β-catenin signaling in an intact organism with physiological levels of upstream Wnt signaling, something that is not possible in mammalian tissue culture. Dishevelled has been reported to be overexpressed in cancer, so having a live model system in which to observe Dishevelled overexpression dynamics may be useful in that regard even its general biological significance would still need to be tested (Uematsu et al., 2003; Mizutani et al., 2005). The ability to study the function and activity of a Wnt signalosome in a live, intact animal, and perhaps one that lacks a LRP5/6 homolog is very promising, and there exists the potential to learn more about canonical signalosome formation here.

Wnt Ligands and Receptors

The roles of the various *C. elegans* Wnt ligands and receptors in polarizing Wnt signaling in the seam cells have been described (Yamamoto et al., 2011). These data show various combinations of Wnts and receptors polarize the various seam cell divisions in the L1 larva, the division analyzed in this study. Given that these multiple, redundant ligands and receptors are then feeding into Dvl function, and given that Dvls in this system are multiple and partially redundant, it would be very interesting to understand whether these multiple Wnts and receptors may be acting in nonredundant ways that were not obvious before. As with DSH-2, MIG-5, and APR-1, understanding how the seam cells regulate multiple outputs of Wnt signaling in a single cell could inform how canonical Wnt/β-catenin signaling could be controlling multiple outputs that are not immediately apparent.

Concluding Remarks

To conclude, we place the WβA pathway into the context of Wnt/β-catenin signaling as a whole. What does one make of a complicated signaling pathway, which controls ACDs in a nematode, and has areas of obvious mechanistic difference with canonical Wnt/β-catenin signaling pathway? Going forward, studies of WβA pathway function during in vivo ACDs will need to leverage its strengths over its weaknesses to answer questions about Wnt signaling that cannot be answered in fixed tissue or cell culture.

The greatest strength is the live, intact nature of the seam cell system. Cell signaling pathways such as Wnt/β-catenin signaling ultimately function through individual cells and how those individual cells respond to the signal, and there are few opportunities in other model systems to watch how a cell responds to Wnt signaling in real time as is possible in the seam cells. Wnt-driven ACDs have been observed in elegant mammalian cell culture experiments, but even those experiments had some drawbacks (Habib et al., 2013). First, these studies rely on Wnt-protein-coated beads to deliver Wnt to the cells, and it is unclear if this presentation mode mimics in vivo Wnt presentation from a polarized source. Second, Wnt-signaling-component localization data, such as APC or β-catenin, depend on on fixed and stained cells, preventing observation of localization dynamics as cell division progresses. Finally, isolated and cultured embryonic stem cells are necessarily removed from their native embryonic environment, in comparison to signaling in an intact organism. We suggest that Wnt-on-a-bead cell polarity experiments are fertile starting points for asking interesting questions, but that the mechanisms gleaned from these experiments should be followed up in in vivo systems wherever possible. To be sure, one could also argue the extent of conservation of the WβA mechanisms discussed above, so these in vitro experiments could be used to extend the worm model in this manner.

While the *C. elegans* seam cells are not a perfect system, they do offer some particular strengths for the study of Wnt/β-catenin signaling. For one, experiments can be performed in a "native" environment of Wnt signaling, where each cell can be presumed to be receiving a physiological level of stimulation rather than subjected to Wnt ligand overexpression. In addition, stable transgenic lines expressing fluorescent fusions of conserved Wnt/β-catenin signaling components can be utilized, allowing for dynamic observation of Wnt signaling in cells as they divide. The worm SCM promoter also allows for tissue-specific expression of transgenes in the seam cells (Terns et al., 1997), allowing the seam cells to even act as a sort of "in vivo tissue culture" if desired. Finally, although there exist important differences, WβA regulation of SYS-1 levels in the seam cells greatly resembles canonical Wnt signaling regulation of β-catenin. The WβA system is sufficiently conserved, in that the functions of many major players in canonical Wnt signaling have been described in the seam cells, Wnt signaling questions that have evaded the vertebrate community may be able to be answered in *C. elegans*. One especially intriguing question, how Wnt signaling regulates multiple roles of APC in a single cell, may be prohibitively difficult to answer in intestinal crypt stem cells but relatively tractable in the seam cells. At the very minimum, the experiments would be significantly less costly and time-consuming in the seam cells, and therefore more adventurous approaches could be utilized.

The WβA pathway in the seam cells is unlikely to be a perfect facsimile of canonical Wnt/β-catenin signaling. There are places where the two pathways obviously diverge, but there are also places where they are fairly similar. With those important differences in mind, careful examination of the WβA pathway has the possibility to answer questions about Wnt signaling, especially the role of Wnt in cell polarity and ACD, which have not been possible so far in vertebrate systems. The genetic strengths of the seam cell system, such as ability to make tissue-specific transgenes as well as relative tolerance to genetic perturbations, may even allow for discovery of novel modulators of Wnt signaling. However, in the event that WβA as a whole turns out to be nematode-specific, this pathway can tell inform the possibilities that exist within conserved Wnt signaling components, which could still result in clinical applications that modulate Wnt signaling. Additionally, there still remains the possibility that a cryptic population of stem cells in the human body is dividing asymmetrically in response to Wnt/β-catenin signaling in a manner highly reminiscent of WβA.

ACKNOWLEDGMENTS

The authors would like to thank the Phillips lab and Douglas Houston for helpful comments on the manuscript. This work was supported by the National Science Foundation (grant number IOS-1456941) and the National Institute of General Medical Sciences (grant number 1R01GM114007-01) to B.T.P.

REFERENCES

Aoki, K., Taketo, M.M., 2007. Adenomatous polyposis coli (APC): a multi-functional tumor suppressor gene. J. Cell Sci. 120, 3327–3335.

Askham, J.M., Moncur, P., Markham, A.F., Morrison, E.E., 2000. Regulation and function of the interaction between the APC tumour suppressor protein and EB1. Oncogene 19, 1950–1958.

Baldwin, A.T., Phillips, B.T., 2014. The tumor suppressor APC differentially regulates multiple beta-catenins through the function of Axin and CKIalpha during *C. elegans* asymmetric stem cell divisions. J. Cell Sci. 127, 2771–2781.

Baldwin, A.T., Clemons, A.M., Phillips, B.T., March 1, 2016. Unique and redundant beta-catenin regulatory roles of two Dishevelled paralogs during *C. elegans* asymmetric cell division. J. Cell Sci. 129 (5), 983–993.

Banerjee, D., Chen, X., Lin, S.Y., Slack, F.J., 2010. kin-19/casein kinase Ialpha has dual functions in regulating asymmetric division and terminal differentiation in *C. elegans* epidermal stem cells. Cell Cycle 9, 4748–4765.

Barker, N., January 2014. Adult intestinal stem cells: critical drivers of epithelial homeostasis and regeneration. Nat. Rev. Mol. Cell. Biol. 15 (1), 19–33. http://dx.doi.org/10.1038/nrm3721.

Barker, N., Ridgway, R.A., van Es, J.H., van de Wetering, M., Begthel, H., van den Born, M., Danenberg, E., Clarke, A.R., Sansom, O.J., Clevers, H., 2009. Crypt stem cells as the cells-of-origin of intestinal cancer. Nature 457, 608–611.

Bei, Y., Hogan, J., Berkowitz, L.A., Soto, M., Rocheleau, C.E., Pang, K.M., Collins, J., Mello, C.C., 2002. SRC-1 and Wnt signaling act together to specify endoderm and to control cleavage orientation in early *C. elegans* embryos. Dev. Cell 3, 113–125.

Bellis, J., Duluc, I., Romagnolo, B., Perret, C., Faux, M.C., Dujardin, D., Formstone, C., Lightowler, S., Ramsay, R.G., Freund, J.N., De Mey, J.R., 2012. The tumor suppressor Apc controls planar cell polarities central to gut homeostasis. J. Cell Biol. 198, 331–341.

Betschinger, J., Mechtler, K., Knoblich, J.A., 2003. The Par complex directs asymmetric cell division by phosphorylating the cytoskeletal protein Lgl. Nature 422, 326–330.

Bianconi, E., Piovesan, A., Facchin, F., Beraudi, A., Casadei, R., Frabetti, F., Vitale, L., Pelleri, M.C., Tassani, S., Piva, F., Perez-Amodio, S., Strippoli, P., Canaider, S., 2013. An estimation of the number of cells in the human body. Ann. Hum. Biol. 40, 463–471.

Bielen, H., Houart, C., 2014. The Wnt cries many: wnt regulation of neurogenesis through tissue patterning, proliferation, and asymmetric cell division. Dev. Neurobiol. 74, 772–780.

Bilić, J., Huang, Y.-L., Davidson, G., Zimmermann, T., Cruciat, C.-M., Bienz, M., Niehrs, C., 2007. Wnt induces LRP6 signalosomes and promotes dishevelled-dependent LRP6 phosphorylation. Science 316, 1619–1622.

Boyd, L., Guo, S., Levitan, D., Stinchcomb, D.T., Kemphues, K.J., 1996. PAR-2 is asymmetrically distributed and promotes association of P granules and PAR-1 with the cortex in *C. elegans* embryos. Development 122, 3075–3084.

Brabin, C., Woollard, A., 2012. Finding a niche for seam cells? Worm 1, 0–4.

Carvajal-Gonzalez, J.M., Mlodzik, M., 2014. Mechanisms of planar cell polarity establishment in *Drosophila*. F1000Prime Rep. 6, 98.

Chang, W., Lloyd, C.E., Zarkower, D., 2005. DSH-2 regulates asymmetric cell division in the early *C. elegans* somatic gonad. Mech. Dev. 122, 781–789.

Chesney, M.A., Lam, N., Morgan, D.E., Phillips, B.T., Kimble, J., 2009. *C. elegans* HLH-2/E/ Daughterless controls key regulatory cells during gonadogenesis. Dev. Biol. 331, 14–25.

Clevers, H., Nusse, R., 2012. Wnt/β-catenin signaling and disease. Cell 149, 1192–1205.

Cliffe, A., Hamada, F., Bienz, M., 2003. A role of Dishevelled in relocating Axin to the plasma membrane during wingless signaling. Curr. Biol. 13, 960–966.

Cong, F., Schweizer, L., Varmus, H., 2004. Wnt signals across the plasma membrane to activate the beta-catenin pathway by forming oligomers containing its receptors, Frizzled and LRP. Development 131, 5103–5115.

Costa, M., Draper, B.W., Priess, J.R., 1997. The role of actin filaments in patterning the *Caenorhabditis elegans* Cuticle. Dev. Biol. 184, 373–384.

Crittenden, S.L., Leonhard, K.A., Byrd, D.T., Kimble, J., 2006. Cellular analyses of the mitotic region in the *Caenorhabditis elegans* adult germ line. Mol. Biol. Cell 17, 3051–3061.

Cselenyi, C.S., Jernigan, K.K., Tahinci, E., Thorne, C.A., Lee, L.A., Lee, E., 2008. LRP6 transduces a canonical Wnt signal independently of Axin degradation by inhibiting GSK3's phosphorylation of beta-catenin. Proc. Natl. Acad. Sci. U.S.A. 105, 8032–8037.

Dillman, A.R., Minor, P.J., Sternberg, P.W., 2013. Origin and evolution of dishevelled. G3 (Bethesda Md.) 3, 251–262.

Driever, W., Nusslein-Volhard, C., 1988. The bicoid protein determines position in the *Drosophila* embryo in a concentration-dependent manner. Cell 54, 95–104.

Driever, W., Nüsslein-Volhard, C., 1988. A gradient of *bicoid* protein in *Drosophila* embryos. Cell 54, 83–93.

Eisenmann, D.M., Maloof, J.N., Simske, J.S., Kenyon, C., Kim, S.K., 1998. The beta-catenin homolog BAR-1 and LET-60 Ras coordinately regulate the Hox gene lin-39 during *Caenorhabditis elegans* vulval development. Dev. (Camb. Engl.) 125, 3667–3680.

Eisenmann, D.M., 2011. *C. elegans* seam cells as stem cells: Wnt signaling and casein kinase Iα regulate asymmetric cell divisions in an epidermal progenitor cell type. Cell Cycle (Georget. Tex.) 10, 20−21.

Etemad-Moghadam, B., Guo, S., Kemphues, K.J., 1995. Asymmetrically distributed PAR-3 protein contributes to cell polarity and spindle alignment in early *C. elegans* embryos. Cell 83, 743−752.

Etheridge, S.L., Ray, S., Li, S., Hamblet, N.S., Lijam, N., Tsang, M., Greer, J., Kardos, N., Wang, J., Sussman, D.J., Chen, P., Wynshaw-Boris, A., 2008. Murine dishevelled 3 functions in redundant pathways with dishevelled 1 and 2 in normal cardiac outflow tract, cochlea, and neural tube development. PLoS Genet. 4.

Fiedler, M., Mendoza-Topaz, C., Rutherford, T.J., Mieszczanek, J., Bienz, M., 2011. Dishevelled interacts with the DIX domain polymerization interface of Axin to interfere with its function in down-regulating β-catenin. Proc. Natl. Acad. Sci. U.S.A. 108, 1937−1942.

Fodde, R., Kuipers, J., Rosenberg, C., Smits, R., Kielman, M., Gaspar, C., van Es, J.H., Breukel, C., Wiegant, J., Giles, R.H., Clevers, H., 2001. Mutations in the APC tumour suppressor gene cause chromosomal instability. Nat. Cell Biol. 3, 433−438.

Gao, C., Chen, Y.-G., 2010. Dishevelled: the hub of Wnt signaling. Cell. Signal. 22, 717−727.

Gessert, S., Kühl, M., 2010. The multiple phases and faces of wnt signaling during cardiac differentiation and development. Circ. Res. 107, 186−199.

Gleason, J.E., Eisenmann, D.M., 2010. Wnt signaling controls the stem cell-like asymmetric division of the epithelial seam cells during *C. elegans* larval development. Dev. Biol. 348, 58−66.

Gleason, J.E., Korswagen, H.C., Eisenmann, D.M., 2002. Activation of Wnt signaling bypasses the requirement for RTK/Ras signaling during *C. elegans* vulval induction. Genes Dev. 16, 1281−1290.

Goldstein, B., Hird, S.N., 1996. Specification of the anteroposterior axis in *Caenorhabditis elegans*. Development 122, 1467−1474.

Goldstein, B., Macara, I.G., 2007. The PAR proteins: fundamental players in animal cell polarization. Dev. Cell 13, 609−622.

Gorrepati, L., Thompson, K.W., Eisenmann, D.M., 2013. *C. elegans* GATA factors EGL-18 and ELT-6 function downstream of Wnt signaling to maintain the progenitor fate during larval asymmetric divisions of the seam cells. Development 140, 2093−2102.

Green, R.A., Kaplan, K.B., 2003. Chromosome instability in colorectal tumor cells is associated with defects in microtubule plus-end attachments caused by a dominant mutation in APC. J. Cell Biol. 163, 949−961.

Green, R.A., Wollman, R., Kaplan, K.B., 2005. APC and EB1 function together in mitosis to regulate spindle dynamics and chromosome alignment. Mol. Biol. Cell 16, 4609−4622.

Groden, J., Thliveris, A., Samowitz, W., Carlson, M., Gelbert, L., Albertsen, H., Joslyn, G., Stevens, J., Spirio, L., Robertson, M., Sargeant, L., Krapcho, K., Wolff, E., Burt, R., Hughes, J.P., Warrington, J., McPherson, J., Wasmuth, J., Le Paslier, D., Abderrahim, H., Cohen, D., Leppert, M., White, R., 1991. Identification and characterization of the familial adenomatous polyposis coli gene. Cell 66, 589−600.

Guo, S., Kemphues, K.J., 1995. par-1, a gene required for establishing polarity in *C. elegans* embryos, encodes a putative Ser/Thr kinase that is asymmetrically distributed. Cell 81, 611−620.

Guo, M., Bier, E., Jan, L.Y., Jan, Y.N., 1995. Tramtrack acts downstream of numb to specify distinct daughter cell fates during asymmetric cell divisions in the *Drosophila* PNS. Neuron 14, 913−925.

Guo, M., Jan, L.Y., Jan, Y.N., 1996. Control of daughter cell fates during asymmetric division: interaction of Numb and Notch. Neuron 17, 27−41.

Ha, N.-C., Tonozuka, T., Stamos, J.L., Choi, H.-J., Weis, W.I., 2004. Mechanism of phosphorylation-dependent binding of APC to beta-catenin and its role in beta-catenin degradation. Mol. Cell 15, 511−521.

Habib, S.J., Chen, B.-C., Tsai, F.-C., Anastassiadis, K., Meyer, T., Betzig, E., Nusse, R., 2013. A localized Wnt signal orients asymmetric stem cell division in vitro. Sci. (N.Y.) 339, 1445−1448.

Hardin, J., 2015. Getting to the core of cadherin complex function in *Caenorhabditis elegans*. F1000Res. 4.

Hart, M.J., de los Santos, R., Albert, I.N., Rubinfeld, B., Polakis, P., 1998. Downregulation of beta-catenin by human Axin and its association with the APC tumor suppressor, beta-catenin and GSK3 beta. Curr. Biol. 8, 573−581.

Hawkins, N.C., Ellis, G.C., Bowerman, B., Garriga, G., 2005. MOM-5 frizzled regulates the distribution of DSH-2 to control C. *elegans* asymmetric neuroblast divisions. Dev. Biol. 284, 246−259.

Henderson, B.R., 2000. Nuclear-cytoplasmic shuttling of APC regulates [beta]-catenin subcellular localization and turnover. Nat. Cell Biol. 2, 653−660.

Hikasa, H., Sokol, S.Y., 2013. Wnt signaling in vertebrate axis specification. Cold Spring Harb. Perspect. Biol. 5, a007955.

Hingwing, K., Lee, S., Nykilchuk, L., Walston, T., Hardin, J., Hawkins, N., 2009. CWN-1 functions with DSH-2 to regulate C. *elegans* asymmetric neuroblast division in a beta-catenin independent Wnt pathway. Dev. Biol. 328, 245−256.

Houston, D.W., 2012. Cortical rotation and messenger RNA localization in *Xenopus* axis formation. Wiley Interdiscip. Rev. Dev. Biol. 1, 371−388.

Huang, S., Shetty, P., Robertson, S.M., Lin, R., 2007. Binary cell fate specification during C. *elegans* embryogenesis driven by reiterated reciprocal asymmetry of TCF POP-1 and its coactivator beta-catenin SYS-1. Dev. (Camb. Engl.) 134, 2685−2695.

Huang, X., Tian, E., Xu, Y., Zhang, H., 2009. The C. *elegans* engrailed homolog ceh-16 regulates the self-renewal expansion division of stem cell-like seam cells. Dev. Biol. 333, 337−347.

Huarcaya Najarro, E., Ackley, B.D., May 1, 2013. C. *elegans* fmi-1/flamingo and Wnt pathway components interact genetically to control the anteroposterior neurite growth of the VD GABAergic neurons. Dev. Biol. 377 (1), 224−235.

Hubbard, E.J.A., 2007. The C. *elegans* germ line: a model for stem cell biology. Dev. Dyn. 236, 3343−3357.

Huber, O., Korn, R., McLaughlin, J., Ohsugi, M., Herrmann, B.G., Kemler, R., 1996. Nuclear localization of beta-catenin by interaction with transcription factor LEF-1. Mech. Dev. 59, 3−10.

Hughes, S., Brabin, C., Appleford, P.J., Woollard, A., June 6, 2013. CEH-20/Pbx and UNC-62/Meis function upstream of rnt-1/Runx to regulate asymmetric divisions of the C. *elegans* stem-like seam cells. Biol. Open 2 (7), 718−727.

Ikeda, S., Kishida, S., Yamamoto, H., Murai, H., Koyama, S., Kikuchi, A., 1998. Axin, a negative regulator of the Wnt signaling pathway, forms a complex with GSK-3β and β-catenin and promotes GSK-3β-dependent phosphorylation of β-catenin. EMBO J. 17, 1371−1384.

Itoh, K., Antipova, A., Ratcliffe, M.J., Sokol, S., 2000. Interaction of dishevelled and *Xenopus* Axin-related protein is required for wnt signal transduction. Mol. Cell. Biol. 20, 2228−2238.

Jackson, B.M., Eisenmann, D.M., August 1, 2012. β-Catenin-Dependent wnt signaling in C. *elegans*: teaching an old dog a new trick. Cold Spring Harb. Perspect. Biol. 4 (8), a007948.

Jaeger, J., Martinez-Arias, A., 2009. Getting the measure of positional information. PLoS Biol. 7, e1000081.

Jensen, M., Hoerndli, F.J., Brockie, P.J., Wang, R., Johnson, E., Maxfield, D., Francis, M.M., Madsen, D.M., Maricq, A.V., 2012. Wnt signaling regulates acetylcholine receptor translocation and synaptic plasticity in the adult nervous system. Cell 149, 173−187.

Kaletta, T., Schnabel, H., Schnabel, R., 1997. Binary specification of the embryonic lineage in *Caenorhabditis elegans*. Nature 390, 294−298.

Kawakami, Y., Capdevila, J., Buscher, D., Itoh, T., Rodriguez Esteban, C., Izpisua Belmonte, J.C., 2001. WNT signals control FGF-dependent limb initiation and AER induction in the chick embryo. Cell 104, 891−900.

Kemphues, K.J., Priess, J.R., Morton, D.G., Cheng, N.S., 1988. Identification of genes required for cytoplasmic localization in early *C. elegans* embryos. Cell 52, 311−320.

Khan, N.P., Pandith, A.A., Hussain, M.U., Yousuf, A., Khan, M.S., Wani, K.A., Mudassar, S., 2011. Novelty of Axin 2 and lack of Axin 1 gene mutation in colorectal cancer: a study in Kashmiri population. Mol. Cell Biochem. 355, 149−155.

Kidd 3rd, A.R., Miskowski, J.A., Siegfried, K.R., Sawa, H., Kimble, J., 2005. A beta-catenin identified by functional rather than sequence criteria and its role in Wnt/MAPK signaling. Cell 121, 761−772.

Kim, C.H., Oda, T., Itoh, M., Jiang, D., Artinger, K.B., Chandrasekharappa, S.C., Driever, W., Chitnis, A.B., 2000. Repressor activity of headless/Tcf3 is essential for vertebrate head formation. Nature 407, 913−916.

Kim, S.E., Huang, H., Zhao, M., Zhang, X., Zhang, A., Semonov, M.V., MacDonald, B.T., Zhang, X., Garcia Abreu, J., Peng, L., He, X., 2013. Wnt stabilization of beta-catenin reveals principles for morphogen receptor-scaffold assemblies. Science 340, 867−870.

King, R.S., Maiden, S.L., Hawkins, N.C., Kidd 3rd, A.R., Kimble, J., Hardin, J., Walston, T.D., 2009. The N- or C-terminal domains of DSH-2 can activate the *C. elegans* Wnt/β-catenin asymmetry pathway. Dev. Biol. 328, 234−244.

Kitagawa, M., Hatakeyama, S., Shirane, M., Matsumoto, M., Ishida, N., Hattori, K., Nakamichi, I., Kikuchi, A., Nakayama, K., Nakayama, K., 1999. An F-box protein, FWD1, mediates ubiquitin-dependent proteolysis of beta-catenin. EMBO J. 18, 2401−2410.

Knoblich, J.A., 2008. Mechanisms of asymmetric stem cell division. Cell 132, 583−597.

Korswagen, H.C., Coudreuse, D.Y.M., Betist, M.C., van de Water, S., Zivkovic, D., Clevers, H.C., 2002. The Axin-like protein PRY-1 is a negative regulator of a canonical Wnt pathway in *C. elegans*. Genes Dev. 16, 1291−1302.

Korswagen, H.C., 2002. Canonical and non-canonical Wnt signaling pathways in *Caenorhabditis elegans*: variations on a common signaling theme. Bioessays 24, 801−810.

Kuchinke, U., Grawe, F., Knust, E., 1998. Control of spindle orientation in *Drosophila* by the par-3-related PDZ-domain protein bazooka. Curr. Biol. 8, 1357−1365.

Lam, A.K., Phillips, B.T., 2017. Wnt signaling polarizes *C. elegans* asymmetric cell divisions during development. Results Probl. Cell Differ. 61, 83−114.

Lander, A.D., Kimble, J., Clevers, H., Fuchs, E., Montarras, D., Buckingham, M., Calof, A.L., Trumpp, A., Oskarsson, T., 2012. What does the concept of the stem cell niche really mean today? BMC Biol. 10, 19.

Lee, E., Salic, A., Krüger, R., Heinrich, R., Kirschner, M.W., 2003. The roles of APC and Axin derived from experimental and theoretical analysis of the wnt pathway. PLoS Biol. 1.

Lewis, W.H., 1904. Experimental studies on the development of the eye in amphibia. I. On the origin of the lens. *Rana palustris*. Am. J. Anat. 3, 505−536.

Li, V.S.W., Ng, S.S., Boersema, P.J., Low, T.Y., Karthaus, W.R., Gerlach, J.P., Mohammed, S., Heck, A.J., Maurice, M.M., Mahmoudi, T., Clevers, H., 2012. Wnt signaling through inhibition of β-catenin degradation in an intact Axin1 complex. Cell 149, 1245−1256.

Lin, R., Thompson, S., Priess, J.R., 1995. pop-1 encodes an HMG box protein required for the specification of a mesoderm precursor in early *C. elegans* embryos. Cell 83, 599−609.

Lin, R., Hill, R.J., Priess, J.R., 1998. POP-1 and anterior-posterior fate decisions in *C. elegans* embryos. Cell 92, 229−239.

Liu, Z., Kirch, S., Ambros, V., 1995. The *Caenorhabditis elegans* heterochronic gene pathway controls stage-specific transcription of collagen genes. Development 121, 2471−2478.

Liu, C., Kato, Y., Zhang, Z., Do, V.M., Yankner, B.A., He, X., 1999. β-Trcp couples β-catenin phosphorylation-degradation and regulates *Xenopus* axis formation. Proc. Natl. Acad. Sci. U.S.A. 96, 6273−6278.

Liu, W., Dong, X., Mai, M., Seelan, R.S., Taniguchi, K., Krishnadath, K.K., Halling, K.C., Cunningham, J.M., Qian, C., Christensen, E., Roche, P.C., Smith, D.I., Thibodeau, S.N., 2000. Mutations in AXIN2 cause colorectal cancer with defective mismatch repair by activating [beta]-catenin/TCF signalling. Nat. Genet. 26, 146−147.

Liu, C., Li, Y., Semenov, M., Han, C., Baeg, G.H., Tan, Y., Zhang, Z., Lin, X., He, X., 2002. Control of beta-catenin phosphorylation/degradation by a dual-kinase mechanism. Cell 108, 837−847.

Liu, J., Phillips, B.T., Amaya, M.F., Kimble, J., Xu, W., 2008. The *C. elegans* SYS-1 protein is a bona fide beta-catenin. Dev. Cell 14, 751−761.

Lo, M.-C., Gay, F., Odom, R., Shi, Y., Lin, R., 2004. Phosphorylation by the β-catenin/MAPK complex promotes 14-3-3-mediated nuclear export of TCF/POP-1 in signal-responsive cells in *C. elegans*. Cell 117, 95−106.

Lopez-Garcia, C., Klein, A.M., Simons, B.D., Winton, D.J., 2010. Intestinal stem cell replacement follows a pattern of neutral drift. Science 330, 822−825.

MacDonald, B.T., Tamai, K., He, X., 2009. Wnt/β-Catenin signaling: components, mechanisms, and diseases. Dev. Cell 17, 9−26.

Maduro, M.F., Meneghini, M.D., Bowerman, B., Broitman-Maduro, G., Rothman, J.H., 2001. Restriction of mesendoderm to a single blastomere by the combined action of SKN-1 and a GSK-3beta homolog is mediated by MED-1 and -2 in *C. elegans*. Mol. Cell 7, 475−485.

Maloof, J.N., Whangbo, J., Harris, J.M., Jongeward, G.D., Kenyon, C., 1999. A Wnt signaling pathway controls hox gene expression and neuroblast migration in *C. elegans*. Dev. (Camb. Engl.) 126, 37−49.

McCrea, P.D., Brieher, W.M., Gumbiner, B.M., 1993. Induction of a secondary body axis in *Xenopus* by antibodies to beta-catenin. J. Cell Biol. 123, 477−484.

Mendoza-Topaz, C., Mieszczanek, J., Bienz, M., 2011. The Adenomatous polyposis coli tumour suppressor is essential for Axin complex assembly and function and opposes Axin's interaction with Dishevelled. Open Biol. 1.

Meneghini, M.D., Ishitani, T., Carter, J.C., Hisamoto, N., Ninomiya-Tsuji, J., Thorpe, C.J., Hamill, D.R., Matsumoto, K., Bowerman, B., 1999. MAP kinase and Wnt pathways converge to downregulate an HMG-domain repressor in *Caenorhabditis elegans*. Nature 399, 793−797.

Mentink, R.A., Middelkoop, T.C., Rella, L., Ji, N., Tang, C.Y., Betist, M.C., van Oudenaarden, A., Korswagen, H.C., 2014. Cell intrinsic modulation of Wnt signaling controls neuroblast migration in *C. elegans*. Dev. Cell 31, 188−201.

Mila, D., Calderon, A., Baldwin, A.T., Moore, K.M., Watson, M., Phillips, B.T., Putzke, A.P., 2015. Asymmetric wnt pathway signaling facilitates stem cell-like divisions via the nonreceptor tyrosine kinase FRK-1 in *Caenorhabditis elegans*. Genetics 201, 1047−1060.

Mimori-Kiyosue, Y., Shiina, N., Tsukita, S., 2000. Adenomatous polyposis coli (APC) protein moves along microtubules and concentrates at their growing ends in epithelial cells. J. Cell Biol. 148, 505−518.

Miskowski, J., Li, Y., Kimble, J., 2001. The sys-1 gene and sexual dimorphism during gonado-genesis in *Caenorhabditis elegans*. Dev. Biol. 230, 61−73.

Mizumoto, K., Sawa, H., 2007a. Cortical β-catenin and APC regulate asymmetric nuclear β-catenin localization during asymmetric cell division in *C. elegans*. Dev. Cell 12, 287−299.

Mizumoto, K., Sawa, H., 2007b. Two βs or not two βs: regulation of asymmetric division by β-catenin. Trends Cell Biol. 17, 465−473.

Mizutani, K., Miyamoto, S., Nagahata, T., Konishi, N., Emi, M., Onda, M., 2005. Upregulation and overexpression of DVL1, the human counterpart of the *Drosophila* dishevelled gene, in prostate cancer. Tumori 91, 546−551.

Molenaar, M., van de Wetering, M., Oosterwegel, M., Peterson-Maduro, J., Godsave, S., Korinek, V., Roose, J., Destrée, O., Clevers, H., 1996. XTcf-3 transcription factor mediates b-catenin-induced axis formation in *Xenopus* embryos. Cell 86, 391−399.

Morin, P.J., Sparks, A.B., Korinek, V., Barker, N., Clevers, H., Vogelstein, B., Kinzler, K.W., 1997. Activation of beta-catenin-Tcf signaling in colon cancer by mutations in beta-catenin or APC. Science 275, 1787−1790.

Morrison, S.J., Kimble, J., 2006. Asymmetric and symmetric stem-cell divisions in development and cancer. Nature 441, 1068−1074.

Motegi, F., Sugimoto, A., 2006. Sequential functioning of the ECT-2 RhoGEF, RHO-1 and CDC-42 establishes cell polarity in *Caenorhabditis elegans* embryos. Nat. Cell Biol. 8, 978−985.

Munemitsu, S., Souza, B., Müller, O., Albert, I., Rubinfeld, B., Polakis, P., 1994. The APC gene product associates with microtubules in vivo and promotes their assembly in vitro. Cancer Res. 54, 3676−3681.

Munemitsu, S., Albert, I., Souza, B., Rubinfeld, B., Polakis, P., 1995. Regulation of intracellular beta-catenin levels by the adenomatous polyposis coli (APC) tumor-suppressor protein. Proc. Natl. Acad. Sci. U.S.A. 92, 3046−3050.

Nakamura, T., Hamada, F., Ishidate, T., Anai, K., Kawahara, K., Toyoshima, K., Akiyama, T., 1998. Axin, an inhibitor of the Wnt signalling pathway, interacts with beta-catenin, GSK-3beta and APC and reduces the beta-catenin level. Genes Cells 3, 395−403.

Nakamura, M., Zhou, X.Z., Lu, K.P., 2001. Critical role for the EB1 and APC interaction in the regulation of microtubule polymerization. Curr. Biol. 11, 1062−1067.

Nakamura, K., Kim, S., Ishidate, T., Bei, Y., Pang, K., Shirayama, M., Trzepacz, C., Brownell, D.R., Mello, C.C., 2005. Wnt signaling drives WRM-1/beta-catenin asymmetries in early *C. elegans* embryos. Genes Dev. 19, 1749−1754.

Nance, J., 2014. Getting to know your neighbor: cell polarization in early embryos. J. Cell Biol. 206, 823−832.

Natarajan, L., Witwer, N.E., Eisenmann, D.M., 2001. The divergent *Caenorhabditis elegans* beta-catenin proteins BAR-1, WRM-1 and HMP-2 make distinct protein interactions but retain functional redundancy in vivo. Genetics 159, 159−172.

Neufeld, K.L., Nix, D.A., Bogerd, H., Kang, Y., Beckerle, M.C., Cullen, B.R., White, R.L., 2000. Adenomatous polyposis coli protein contains two nuclear export signals and shuttles between the nucleus and cytoplasm. Proc. Natl. Acad. Sci. U.S.A. 97, 12085−12090.

Neumüller, R.A., Knoblich, J.A., 2009. Dividing cellular asymmetry: asymmetric cell division and its implications for stem cells and cancer. Genes Dev. 23, 2675−2699.

Oh, J., Lee, Y.D., Wagers, A.J., 2014. Stem cell aging: mechanisms, regulators and therapeutic opportunities. Nat. Med. 20, 870−880.

Ohshiro, T., Yagami, T., Zhang, C., Matsuzaki, F., 2000. Role of cortical tumour-suppressor proteins in asymmetric division of *Drosophila* neuroblast. Nature 408, 593−596.

Okamoto, M., Inoue, K., Iwamura, H., Terashima, K., Soya, H., Asashima, M., Kuwabara, T., 2011. Reduction in paracrine Wnt3 factors during aging causes impaired adult neurogenesis. FASEB J. 25, 3570–3582.

Orford, K., Crockett, C., Jensen, J.P., Weissman, A.M., Byers, S.W., 1997. Serine phosphorylation-regulated ubiquitination and degradation of β-catenin. J. Biol. Chem. 272, 24735–24738.

Ozawa, M., Baribault, H., Kemler, R., 1989. The cytoplasmic domain of the cell adhesion molecule uvomorulin associates with three independent proteins structurally related in different species. EMBO J. 8, 1711–1717.

Park, J.Y., Park, W.S., Nam, S.W., Kim, S.Y., Lee, S.H., Yoo, N.J., Lee, J.Y., Park, C.K., 2005. Mutations of beta-catenin and AXIN I genes are a late event in human hepatocellular carcinogenesis. Liver Int. 25, 70–76.

Peng, C.Y., Manning, L., Albertson, R., Doe, C.Q., 2000. The tumour-suppressor genes lgl and dlg regulate basal protein targeting in *Drosophila* neuroblasts. Nature 408, 596–600.

Peters, J.M., McKay, R.M., McKay, J.P., Graff, J.M., 1999. Casein kinase I transduces Wnt signals. Nature 401, 345–350.

Petronczki, M., Knoblich, J.A., 2001. DmPAR-6 directs epithelial polarity and asymmetric cell division of neuroblasts in *Drosophila*. Nat. Cell Biol. 3, 43–49.

Pfister, A.S., Hadjihannas, M.V., Röhrig, W., Schambony, A., Behrens, J., 2012. Amer2 protein interacts with EB1 protein and adenomatous polyposis coli (APC) and controls microtubule stability and cell migration. J. Biol. Chem. 287, 35333–35340.

Phillips, B.T., Kimble, J., 2009. A new look at TCF and beta-catenin through the lens of a divergent *C. elegans* Wnt pathway. Dev. Cell 17, 27–34.

Phillips, B.T., Kidd 3rd, A.R., King, R., Hardin, J., Kimble, J., 2007. Reciprocal asymmetry of SYS-1/beta-catenin and POP-1/TCF controls asymmetric divisions in *Caenorhabditis elegans*. Proc. Natl. Acad. Sci. U.S.A. 104, 3231–3236.

Price, M.A., 2006. CKI, there's more than one: casein kinase I family members in Wnt and Hedgehog signaling. Genes Dev. 20, 399–410.

Punchihewa, C., Ferreira, A.M., Cassell, R., Rodrigues, P., Fujii, N., 2009. Sequence requirement and subtype specificity in the high-affinity interaction between human frizzled and dishevelled proteins. Protein Sci. 18, 994–1002.

Quyn, A.J., Appleton, P.L., Carey, F.A., Steele, R.J., Barker, N., Clevers, H., Ridgway, R.A., Sansom, O.J., Nathke, I.S., 2010. Spindle orientation bias in gut epithelial stem cell compartments is lost in precancerous tissue. Cell Stem Cell 6, 175–181.

Reguart, N., He, B., Taron, M., You, L., Jablons, D.M., Rosell, R., 2005. The role of Wnt signaling in cancer and stem cells. Future Oncol. 1, 787–797.

Reilein, A., Nelson, W.J., 2005. APC is a component of an organizing template for cortical microtubule networks. Nat. Cell Biol. 7, 463–473.

Ren, H., Zhang, H., 2010. Wnt signaling controls temporal identities of seam cells in *Caenorhabditis elegans*. Dev. Biol. 345, 144–155.

Rhyu, M.S., Jan, L.Y., Jan, Y.N., 1994. Asymmetric distribution of numb protein during division of the sensory organ precursor cell confers distinct fates to daughter cells. Cell 76, 477–491.

Rocheleau, C.E., Yasuda, J., Shin, T.H., Lin, R., Sawa, H., Okano, H., Priess, J.R., Davis, R.J., Mello, C.C., 1999. WRM-1 activates the LIT-1 protein kinase to transduce anterior/posterior polarity signals in *C. elegans*. Cell 97, 717–726.

Rogers, K.W., Schier, A.F., 2011. Morphogen gradients: from generation to interpretation. Annu. Rev. Cell Dev. Biol. 27, 377–407.

Rosin-Arbesfeld, R., Townsley, F., Bienz, M., 2000. The APC tumour suppressor has a nuclear export function. Nature 406, 1009–1012.

Rosin-Arbesfeld, R., Cliffe, A., Brabletz, T., Bienz, M., 2003. Nuclear export of the APC tumour suppressor controls β-catenin function in transcription. EMBO J. 22, 1101−1113.

Rubinfeld, B., Souza, B., Albert, I., Munemitsu, S., Polakis, P., 1995. The APC protein and E-cadherin form similar but independent complexes with alpha-catenin, beta-catenin, and plakoglobin. J. Biol. Chem. 270, 5549−5555.

Rubinfeld, B., Albert, I., Porfiri, E., Fiol, C., Munemitsu, S., Polakis, P., 1996. Binding of GSK3beta to the APC-beta-catenin complex and regulation of complex assembly. Sci. (N.Y.) 272, 1023−1026.

Rubinfeld, B., Tice, D.A., Polakis, P., 2001. Axin-dependent phosphorylation of the adenomatous polyposis coli protein mediated by casein kinase 1ε. J. Biol. Chem. 276, 39037−39045.

Sanchez-Alvarez, L., Visanuvimol, J., McEwan, A., Su, A., Imai, J.H., Colavita, A., 2011. VANG-1 and PRKL-1 cooperate to negatively regulate neurite formation in *Caenorhabditis elegans*. PLoS Genet. 7.

Sawa, H., Korswagen, H.C., 2013. Wnt signaling in *C. elegans*. WormBook 1−30.

Sawa, H., 2012. Control of cell polarity and asymmetric division in *C. elegans*. Curr. Top. Dev. Biol. 101, 55−76.

Schaefer, M., Shevchenko, A., Shevchenko, A., Knoblich, J.A., 2000. A protein complex containing inscuteable and the Gα-binding protein Pins orients asymmetric cell divisions in *Drosophila*. Curr. Biol. 10, 353−362.

Schlesinger, A., Shelton, C.A., Maloof, J.N., Meneghini, M., Bowerman, B., 1999. Wnt pathway components orient a mitotic spindle in the early *Caenorhabditis elegans* embryo without requiring gene transcription in the responding cell. Genes Dev. 13, 2028−2038.

Schneider, S., Steinbeisser, H., Warga, R.M., Hausen, P., 1996. Beta-catenin translocation into nuclei demarcates the dorsalizing centers in frog and fish embryos. Mech. Dev. 57, 191−198.

Schonegg, S., Hyman, A.A., 2006. CDC-42 and RHO-1 coordinate acto-myosin contractility and PAR protein localization during polarity establishment in *C. elegans* embryos. Development 133, 3507−3516.

Schwarz-Romond, T., Fiedler, M., Shibata, N., Butler, P.J., Kikuchi, A., Higuchi, Y., Bienz, M., 2007. The DIX domain of dishevelled confers Wnt signaling by dynamic polymerization. Nat. Struct. Mol. Biol. 14, 484−492.

Seib, D.R.M., Corsini, N.S., Ellwanger, K., Plaas, C., Mateos, A., Pitzer, C., Niehrs, C., Celikel, T., Martin-Villalba, A., 2013. Loss of Dickkopf-1 restores neurogenesis in old age and counteracts cognitive decline. Cell Stem Cell 12, 204−214.

Shin, T.H., Yasuda, J., Rocheleau, C.E., Lin, R., Soto, M., Bei, Y., Davis, R.J., Mello, C.C., 1999. MOM-4, a MAP kinase kinase kinase-related protein, activates WRM-1/LIT-1 kinase to transduce anterior/posterior polarity signals in *C. elegans*. Mol. Cell 4, 275−280.

Siegfried, K.R., Kidd 3rd, A.R., Chesney, M.A., Kimble, J., 2004. The sys-1 and sys-3 genes cooperate with Wnt signaling to establish the proximal-distal axis of the *Caenorhabditis elegans* gonad. Genetics 166, 171−186.

Singh, R.N., Sulston, J.E., 1978. Some observations on moulting in *Caenorhabditis Elegans*. Nematologica 24, 63−71.

Smith, K.J., Levy, D.B., Maupin, P., Pollard, T.D., Vogelstein, B., Kinzler, K.W., 1994. Wild-type but not mutant APC associates with the microtubule cytoskeleton. Cancer Res. 54, 3672−3675.

Sokol, S.Y., 1999. Wnt signaling and dorso-ventral axis specification in vertebrates. Curr. Opin. Genet. Dev. 9, 405−410.

Sokol, S.Y., 2015. Spatial and temporal aspects of Wnt signaling and planar cell polarity during vertebrate embryonic development. Semin. Cell Dev. Biol. 42, 78−85.

Song, X., Wang, S., Li, L., 2014. New insights into the regulation of Axin function in canonical Wnt signaling pathway. Protein Cell 5, 186−193.

Spemann, H., Mangold, H., 1923. 1923 induction of embryonic primordia by implantation of organizers from a different species. Int. J. Dev. Biol. 45, 13−38.

Spink, K.E., Polakis, P., Weis, W.I., 2000. Structural basis of the Axin−adenomatous polyposis coli interaction. EMBO J. 19, 2270−2279.

Stamos, J.L., Weis, W.I., 2013. The β-catenin destruction complex. Cold Spring Harb. Perspect. Biol. 5.

Struhl, G., Struhl, K., Macdonald, P.M., 1989. The gradient morphogen bicoid is a concentration-dependent transcriptional activator. Cell 57, 1259−1273.

Struhl, G., 1989. Morphogen gradients and the control of body pattern in insect embryos. Ciba Found. Symp. 144, 65−86 discussion 86−91, 92−68.

Sugioka, K., Mizumoto, K., Sawa, H., 2011. Wnt regulates spindle asymmetry to generate asymmetric nuclear beta-catenin in C. elegans. Cell 146, 942−954.

Sulston, J.E., Horvitz, H.R., 1977. Post-embryonic cell lineages of the nematode, Caenorhabditis elegans. Dev. Biol. 56, 110−156.

Sulston, J.E., Schierenberg, E., White, J.G., Thomson, J.N., 1983. The embryonic cell lineage of the nematode Caenorhabditis elegans. Dev. Biol. 100, 64−119.

Taelman, V.F., Dobrowolski, R., Plouhinec, J.L., Fuentealba, L.C., Vorwald, P.P., Gumper, I., Sabatini, D.D., De Robertis, E.M., 2010. Wnt signaling requires sequestration of glycogen synthase kinase 3 inside multivesicular endosomes. Cell 143, 1136−1148.

Takeshita, H., Sawa, H., 2005. Asymmetric cortical and nuclear localizations of WRM-1/beta-catenin during asymmetric cell division in C. elegans. Genes Dev. 19, 1743−1748.

Tauriello, D.V.F., Jordens, I., Kirchner, K., Slootstra, J.W., Kruitwagen, T., Bouwman, B.A.M., Noutsou, M., Rüdiger, S.G.D., Schwamborn, K., Schambony, A., Maurice, M.M., 2012. Wnt/β-catenin signaling requires interaction of the dishevelled DEP domain and C terminus with a discontinuous motif in Frizzled. Proc. Natl. Acad. Sci. U.S.A. 109, E812−E820.

ten Berge, D., Kurek, D., Blauwkamp, T., Koole, W., Maas, A., Eroglu, E., Siu, R.K., Nusse, R., 2011. Embryonic stem cells require Wnt proteins to prevent differentiation to epiblast stem cells. Nat. Cell Biol. 13, 1070−1075.

Terns, R.M., Kroll-Conner, P., Zhu, J., Chung, S., Rothman, J.H., 1997. A deficiency screen for zygotic loci required for establishment and patterning of the epidermis in Caenorhabditis elegans. Genetics 146, 185−206.

Tetsu, O., McCormick, F., 1999. Beta-catenin regulates expression of cyclin D1 in colon carcinoma cells. Nature 398, 422−426.

Thein, M.C., McCormack, G., Winter, A.D., Johnstone, I.L., Shoemaker, C.B., Page, A.P., 2003. Caenorhabditis elegans exoskeleton collagen COL-19: an adult-specific marker for collagen modification and assembly, and the analysis of organismal morphology. Dev. Dyn. 226, 523−539.

Tickle, C., Towers, M., 2017. Sonic hedgehog signaling in limb development. Front. Cell Dev. Biol. 5, 14.

Uematsu, K., Kanazawa, S., You, L., He, B., Xu, Z., Li, K., Peterlin, B.M., McCormick, F., Jablons, D.M., 2003. Wnt pathway activation in mesothelioma: evidence of dishevelled overexpression and transcriptional activity of beta-catenin. Cancer Res. 63, 4547−4551.

Uemura, T., Shepherd, S., Ackerman, L., Jan, L.Y., Jan, Y.N., 1989. numb, a gene required in determination of cell fate during sensory organ formation in Drosophila embryos. Cell 58, 349−360.

Valvezan, A.J., Zhang, F., Diehl, J.A., Klein, P.S., 2012. Adenomatous polyposis coli (APC) regulates multiple signaling pathways by enhancing glycogen synthase kinase-3 (GSK-3) activity. J. Biol. Chem. 287, 3823–3832.

van Noort, M., Meeldijk, J., van der Zee, R., Destree, O., Clevers, H., 2002. Wnt signaling controls the phosphorylation status of beta-catenin. J. Biol. Chem. 277, 17901–17905.

Vora, S., Phillips, B.T., 2015. Centrosome-associated degradation limits beta-catenin inheritance by daughter cells after asymmetric division. Curr. Biol. 25, 1005–1016.

Vora, S.M., Phillips, B.T., 2016. The benefits of local depletion: the centrosome as a scaffold for ubiquitin-proteasome-mediated degradation. Cell Cycle 15, 2124–2134.

Wallingford, J.B., Mitchell, B., 2011. Strange as it may seem: the many links between Wnt signaling, planar cell polarity, and cilia. Genes Dev. 25, 201–213.

Walston, T., Tuskey, C., Edgar, L., Hawkins, N., Ellis, G., Bowerman, B., Wood, W., Hardin, J., 2004. Multiple Wnt signaling pathways converge to orient the mitotic spindle in early C. elegans embryos. Dev. Cell 7, 831–841.

Walston, T., Guo, C., Proenca, R., Wu, M., Herman, M., Hardin, J., Hedgecock, E., 2006. mig-5/ Dsh controls cell fate determination and cell migration in C. elegans. Dev. Biol. 298, 485–497.

Watts, J.L., Etemad-Moghadam, B., Guo, S., Boyd, L., Draper, B.W., Mello, C.C., Priess, J.R., Kemphues, K.J., 1996. par-6, a gene involved in the establishment of asymmetry in early C. elegans embryos, mediates the asymmetric localization of PAR-3. Development 122, 3133–3140.

Webster, M.T., Rozycka, M., Sara, E., Davis, E., Smalley, M., Young, N., Dale, T.C., Wooster, R., 2000. Sequence variants of the Axin gene in breast, colon, and other cancers: an analysis of mutations that interfere with GSK3 binding. Genes Chromosom. Cancer 28, 443–453.

Wieschaus, E., Riggleman, R., 1987. Autonomous requirements for the segment polarity gene armadillo during Drosophila embryogenesis. Cell 49, 177–184.

Wildwater, M., Sander, N., de Vreede, G., van den Heuvel, S., 2011. Cell shape and Wnt signaling redundantly control the division axis of C. elegans epithelial stem cells. Dev. (Camb. Engl.) 138, 4375–4385.

Wodarz, A., Ramrath, A., Grimm, A., Knust, E., 2000. Drosophila atypical protein kinase C associates with Bazooka and controls polarity of epithelia and neuroblasts. J. Cell Biol. 150, 1361–1374.

Wolpert, L., 1968. The {French flag} problem: a contribution to the discussion on pattern development and regulation. In: Waddington, C.H. (Ed.), The Origin of Life: Toward a Theoretical Biology, pp. 125–133.

Wolpert, L., 1989. Positional information revisited. Development 107 (Suppl.), 3–12.

Wong, H.C., Mao, J., Nguyen, J.T., Srinivas, S., Zhang, W., Liu, B., Li, L., Wu, D., Zheng, J., 2000. Structural basis of the recognition of the dishevelled DEP domain in the Wnt signaling pathway. Nat. Struct. Biol. 7, 1178–1184.

Wong, H.-C., Bourdelas, A., Krauss, A., Lee, H.-J., Shao, Y., Wu, D., Mlodzik, M., Shi, D.-L., Zheng, J., 2003. Direct binding of the PDZ domain of dishevelled to a conserved internal sequence in the C-terminal region of Frizzled. Mol. Cell 12, 1251–1260.

Wu, M., Herman, M.A., 2006. A novel noncanonical Wnt pathway is involved in the regulation of the asymmetric B cell division in C. elegans. Dev. Biol. 293, 316–329.

Wu, M., Herman, M.A., 2007. Asymmetric localizations of LIN-17/Fz and MIG-5/Dsh are involved in the asymmetric B cell division in C. elegans. Dev. Biol. 303, 650–662.

Xing, Y., Clements, W.K., Trong, I.L., Hinds, T.R., Stenkamp, R., Kimelman, D., Xu, W., 2004. Crystal structure of a β-catenin/APC complex reveals a critical role for APC phosphorylation in APC function. Mol. Cell 15, 523—533.

Yamamoto, Y., Takeshita, H., Sawa, H., 2011. Multiple Wnts redundantly control polarity orientation in *Caenorhabditis elegans* epithelial stem cells. PLoS Genet. 7, e1002308.

Yang, X.D., Huang, S., Lo, M.C., Mizumoto, K., Sawa, H., Xu, W., Robertson, S., Lin, R., 2011. Distinct and mutually inhibitory binding by two divergent beta-catenins coordinates TCF levels and activity in *C. elegans*. Development 138, 4255—4265.

Yost, C., Torres, M., Miller, J.R., Huang, E., Kimelman, D., Moon, R.T., 1996. The axis-inducing activity, stability, and subcellular distribution of beta-catenin is regulated in *Xenopus* embryos by glycogen synthase kinase 3. Genes Dev. 10, 1443—1454.

Yu, F., Morin, X., Cai, Y., Yang, X., Chia, W., 2000. Analysis of partner of inscuteable, a novel player of *Drosophila* asymmetric divisions, reveals two distinct steps in inscuteable apical localization. Cell 100, 399—409.

Yu, F., Kuo, C.T., Jan, Y.N., 2006. *Drosophila* neuroblast asymmetric cell division: recent advances and implications for stem cell biology. Neuron 51, 13—20.

Zacharias, A.L., Murray, J.I., 2016. Combinatorial decoding of the invariant *C. elegans* embryonic lineage in space and time. Genesis 54, 182—197.

Zacharias, A.L., Walton, T., Preston, E., Murray, J.I., 2015. Quantitative differences in nuclear beta-catenin and TCF pattern embryonic cells in *C. elegans*. PLoS Genet. 11, e1005585.

Zeng, L., Fagotto, F., Zhang, T., Hsu, W., Vasicek, T.J., Perry 3rd, W.L., Lee, J.J., Tilghman, S.M., Gumbiner, B.M., Costantini, F., 1997. The mouse Fused locus encodes Axin, an inhibitor of the Wnt signaling pathway that regulates embryonic axis formation. Cell 90, 181—192.

Zumbrunn, J., Kinoshita, K., Hyman, A.A., Näthke, I.S., 2001. Binding of the adenomatous polyposis coli protein to microtubules increases microtubule stability and is regulated by GSK3β phosphorylation. Curr. Biol. 11, 44—49.

Chapter 4

Cell Polarity in Morphogenesis—Planar Cell Polarity

Noopur Mandrekar[1], Baihao Su[1], Raymond Habas[1,2]
[1]Temple University, Philadelphia, PA, United States; [2]Fox Chase Cancer Center, Philadelphia, PA, United States

INTRODUCTION

During embryonic development, the coordinated and distinct polarization of cells is a key phenomenon for the evolution and establishment of a three-dimension form within a group of cells. This organization provides the basis for establishment of the primary embryonic axis, cell migration, neural tube closure, and tissue morphogenesis. How such planar polarity is initiated and elaborated in static as well as motile cells has remained intensively studied, and while a plethora of molecular players have emerged, our understanding of the mechanisms behind this key event remains somewhat rudimentary. In this chapter, we will review our current understanding of known molecular components and emerging mechanisms that govern planar cell polarity (PCP).

The organization of the group of cells beginning from the early cleavage stages of development in numerous organisms was observed for many centuries. However, early observations with primitive microscopes led to many controversies about how the embryo "morphed" into its precise three-dimensional shapes. One such interesting controversy led to the "preformation" hypothesis of Marcello Malphigi in the late 1600s versus the opposing "epigenesis" hypothesis by Caspar Wolff in the mid 1700s (De Felici and Siracusa, 2000; Witt, 2008). But as better microscopes arose along with advances in molecular biology, an explosion continues into our understanding of the behaviors of cell and the signaling pathways that control such activity. The concept of PCP, which is also referred to as "tissue polarity," was first studied in the retina of the insect, *Oncopeltus fasciatus*. Within the retina, it was observed that ommatidia were all oriented in the same direction across that tissue (Lawrence and Shelton, 1975). Subsequently, such polarization of structures was seen in

Cell Polarity in Development and Disease. https://doi.org/10.1016/B978-0-12-802438-6.00004-8

103

other tissues including hair cells and bristle orientation. This phenomenon was realized to not be limited to insects, as studies have uncovered key roles for PCP proteins in regulating cell migration during gastrulation and neural tube closure, axonal guidance, and neuronal migration and polarization of tissue structures such as inner hair cells and stereocilia (Hale and Strutt, 2015; Davey and Moens, 2017).

Approaches to dissect the formation and control of orientation of such structures were first undertaken in *Drosophila melanogaster* where large-scale genetic screening methodologies were developed to identify early genes required for patterning of the early embryo. The visual ease by which the orientation of polarized structures such as hair bristles could be identified allowed for the successful identification of mutants that contained perturbations in orientation of these structures (Lawrence et al., 2002; Tree et al., 2002a). Interestingly, these screens have identified components of two core conserved signaling pathways that are termed the noncanonical Wnt or Wnt/PCP pathway and the Fat/Dachsous/Four-jointed pathway. In this chapter, we will limit our focus to the role of the Wnt pathway.

The genetic and subsequent molecular studies in *Drosophila* led to identification of key proteins that were involved originally in patterning the early *Drosophila* embryo that belonged to the canonical Wnt signaling pathway that regulated cell fate determination. However, refined screening approaches examining planar orientation of structures within epithelial structures and the eye also uncovered these canonical Wnt components such as Dishevelled (Dvl) (Theisen et al., 1994), Frizzled (Fz) (Vinson and Adler, 1987), Van Gogh (Taylor et al., 1998), and Flamingo (Usui et al., 1999). These studies together uncovered additional roles for the Wnt signaling pathway in the control of planar polarization that was independent of the canonical Wnt pathway, and this branch of signaling was termed as the noncanonical Wnt signaling pathway. Mutations in any of these core proteins result in disorganization of ommatidia and misorientation of bristles though the tissues were fully specified suggesting a transcriptional independent role of Wnt signaling for PCP. It is important to note that Wnt components such as Frizzed and Dvl were identified as key component for the noncanonical pathway; the studies in *Drosophila* failed to demonstrate any of the known four *Drosophila* Wnts which may serve as signaling ligands for this pathway. Hence for many years, there was some confusion as to whether the Wnt/PCP pathway was indeed manifested as the noncanonical Wnt signaling pathway biochemically and developmentally characterized in vertebrates as required for gastrulation (Mlodzik, 2002; Seifert and Mlodzik, 2007). Recent studies have now begun to clarify this, see below, and it is now generally accepted that the two pathways are likely the same or at least share a large overlapping cohort of components.

During early development, in both invertebrates and vertebrates, the physical body plan of the embryo emerges from a series of dynamic cell migratory events of cells within the primary germ layers, primarily dorsal mesodermal cells. While intrinsic differences are observed in the described types of movements and behavior of cells that occur within the varying phyla, these cell behaviors all require cell polarization, adhesion, and directed migration. The types of cell behaviors have been extensively described in *Drosophila melanogaster, Caenorhabditis elegans*, and numerous vertebrate species including *Xenopus*, zebrafish, chick, urchins, and mouse. Thus, we now have a wealth of information that has been gleaned across the various species, and these studies within the last decade have further defined key molecular components that are required for such cell behaviors.

In vertebrates, gastrulation is a key morphogenetic event that establishes the body axis of the embryo. In frog development, gastrulation is characterized by blastopore closure moving the endoderm and mesoderm on the inner side and below the ectoderm giving rise to archenteron to establish and shape the three primary germ layers (Beetschen, 2001). In zebrafish, epiboly and emboly dominate the gastrulation process. The major morphogenetic movements following internalization are the convergent and extension of the meso-endoderm. Convergent extension is a critical process during gastrulation involving the elongation of mesenchymal cells to form mediolateral (ML) protrusions driving intercalation of cells toward the midline so that the embryo narrows along the ML axis and lengthens along the anterior posterior (AP) axis (Solnica-Krezel and Sepich, 2012). In zebrafish, the mesenchymal cells intercalate during elongation of the dorsal mesoderm and ectoderm (Glickman et al., 2003). Impairment in convergent and extension results in *Xenopus*, zebrafish, and mouse embryos with open neural tube and shortened and wide body axis (Wallingford and Harland, 2002; Wang et al., 2006).

Interestingly, convergent extension is also redeployed during neural tube closure (Park et al., 2006). The convergent extension of the neural plate and dorsal mesoderm are dependent on the PCP pathway (Wallingford and Harland, 2002; Rolo et al., 2009). The core PCP proteins are required for axis elongation in all vertebrate models examined this far. Congenital malformations of the brain also called as neural tube defect arise from the failure of neural tube closure during embryogenesis. These malformations affect the development of the spinal cord and cranium manifesting into spina bifida and anencephaly. PCP mutations have also been linked to human disorders such as Robinow syndrome, scoliosis, and spina bifida (Andre et al., 2012; Wen et al., 2010; Andersen et al., 2016). Gain-of-function and loss-of-function approaches in vertebrate models such as *Xenopus*, zebrafish, and mice have been instrumental in deciphering the essential role of PCP pathway in these events.

In addition to dynamic cell polarization and shape changes that occur during gastrulation, another key cell shape change is apical constriction. During this process, cells at the apical surface, which are usually cuboid or columnar, transform into a trapezoidal or wedge shape that causes the epithelial cell to bend such that the basal surface becomes convex (Lee and Harland, 2007). In the initial stages of *Xenopus* gastrulation, cells in the dorsal marginal zone undergo the process of apical constriction and lengthen along the apicobasal axis, transforming into flask-shaped "bottle cells." These bottle cells initiate the formation of the blastopore lip, allowing cells to internalize (Lee and Harland, 2007). During neural tube apposition and closure, apical construction also plays a vital role. Both F-actin and myosin are known to play a major role in bottle cell formation and efficient apical constriction, while intact microtubules are required for apical constriction, but not elongation, of bottle cells (Lee and Harland, 2007).

THE PLANAR CELL POLARITY PATHWAY

Wnts

PCP pathway is evolutionarily conserved from *Drosophila* to mammals. Wnt proteins have been considered as potential candidates to provide cues for establishment of tissue polarity. In *Drosophila*, the mix-expression of Wingless and Wnt4a leads to reorientation of wing hair polarity suggesting that a Wnt gradient is needed for establishment of PCP in the wing margin (Wu et al., 2013). In zebrafish, genetic studies uncovered that Wnt5a/*pipetail* and Wnt11/ *silberblick* play an important role during convergent extension movements (Kilian et al., 2003; Heisenberg et al., 2000) in addition to other core components of the pathway. In a recent study, during *Xenopus* gastrulation, it was observed that Wnt ligands, Wnt5a, Wnt11, Wnt11b instruct exogenous Prickle3/Vangl2 complex in the epidermis and endogenous Vangl2 in the neuroectoderm to drive PCP in ectodermal cells (Chu and Sokol, 2016).

In the noncanonical Wnt pathway, the Wnt signal is mediated through the Frizzled (Fz) receptor in absence of the Lrp5/6 coreceptor, but there have been conflicting studies implicating a role for Lrp5/6 in regulation of convergent extension (He et al., 2004; Tahinci et al., 2007) (Figs. 4.1 and 4.2). Epistatic studies suggest that Fz2 and Fz7 serve as receptors for Wnt5a and Wnt11 during gastrulation in zebrafish and *Xenopus* (Kilian et al., 2003; Djiane et al., 2000). Wnt11 can regulate the subcellular localization of Fz7 as overexpressed Fz7 is observed localized at the cell membranes during zebrafish gastrulation. While the role of the Fz proteins as receptors has been cemented, a number of studies have identified putative candidate transmembrane proteins as coreceptors. These currently include NRH1, Ryk, PTK7, and ROR2 (Sasai et al., 2004; Lu et al., 2004a,b; Nishita et al., 2006). From the receptor/coreceptor complex, the signal is then transduced to the cytoplasmic phosphoprotein Dvl. At the level of Dvl and using distinct domains, the PDZ and DEP domains have been shown to activate two parallel pathways using biochemical and epistasis analyses. The PDZ domain leads to activation of the small GTPase Rho and this activation of Rho has been shown to be mediated through the Formin (FMN)-homology protein DAAM1 that interacts with the PDZ and DEP domains of Dvl. Activated small

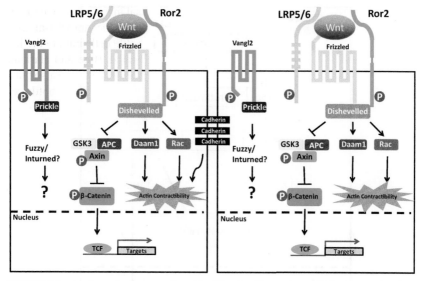

FIGURE 4.1 A schematic representation of the Wnt signaling pathway. (1) In the canonical Wnt signaling pathway, signaling through Frizzled (*Fz*) and coreceptor Lrp5/6 recruits Dishevelled (*Dvl*) to the plasma membrane, which leads to the dissociation of the "destruction complex" that is made up of: *Axin*, adenomatosis polyposis coli (*APC*), glycogen synthesis kinase 3 (*GSK3*), and casein kinase 1α (*CK1α*). Therefore, accumulated *β-catenin* then translocates from cytoplasm into nucleus to bind to T cell factor (*TCF*) and triggers the transcription of the Wnt target genes. (2) In the noncanonical Wnt signaling pathway or planar cellular polarity (PCP) signaling pathway, Wnt signaling causes the formation of two complexes: Fz/Dvl and Vangle2/Prickle. Signaling through Fz/Dvl complex mediates actin contractibility through small GTPase: Rho and Rac. Signaling through Vangle2/Prickle regulates the activity of downstream PCP proteins: Fuzzy and inturned whose functions are still unclear. The PCP is coordinated by the antagonistic interaction of those two complexes within the cells and positive feedback *a*cross the cell junction. Two adjacent cells are shown above.

GTPase Rho then activates the Rho-associated kinase ROCK, ultimately leading to modification of actin cytoskeleton and cytoskeletal rearrangement (Habas et al., 2001; Marlow et al., 2002). The second pathway activates Rac GTPases to stimulate c-Jun N-terminal Kinase (JNK) activity via the DEP domain of Dvl, which plays critical role in modulation of the actin cytoskeleton (Li et al., 1999; Keller et al., 2003). While JNK has been known to play a role in transcriptional regulation, to date it remains unclear whether any transcriptional target of JNK is required for PCP regulation.

Asymmetric Protein Localization: From Flies to Vertebrates

In the PCP pathway in *Drosophila*, studies have uncovered the key role of the core components of the Wnt noncanonical pathway including Fz, Dvl, Strabismus (Vangl in vertebrates), Flamingo (Celsr in mammals), Prickle, and

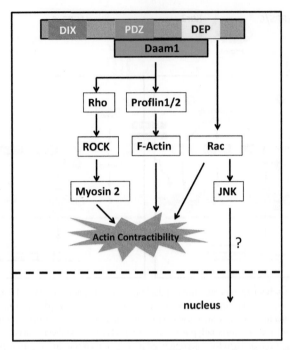

FIGURE 4.2 A schematic representation of the Wnt/planar cell polarity (PCP) signaling pathway. The Wnt/PCP signaling is transduced through *Frizzled* independently of Lrp5/6 to *Dvl* where it splits into two independent branches: (1) Signaling to Rho and proflin1/2 occur through Daam1. The Rho pathway leads to the activation of Rho associate kinase (ROCK) and its target myosin2, which drives actomyosin contractility. The proflin1/2 pathway leads to the changes in polymerization of actin filaments. (2) The activation of Rac requires the DEP domain of *Dvl* which in turn leads to activation of c-Jun N-terminal kinase (*JNK*). The role of JNK in regulating transcription remains unclear.

Diego (Inversin/Diversin in vertebrates). The subcellular localization of the components of noncanonical Wnt pathway reveal that they are recruited uniformly around the subapical cell membranes followed by asymmetric distribution in the polarized epithelia (Gray et al., 2011). Some of these components such as Prickle and Strabismus are localized in the proximal part of the polarized cell membrane of the *Drosophila* wing cells, whereas proteins such as Fz, Dvl, and Diego accumulate in the distal region of the cell membrane, thus communicating polarity information between adjacent cells (Tree et al., 2002b; Axelrod, 2001; Strutt, 2001) (Fig. 4.3). Overexpression or downregulation of any of the core PCP components can perturb the localization of other PCP proteins (Seifert and Mlodzik, 2007). The Fat/Dachsous system includes the cadherins, Fat and Dachsous, and the Golgi protein Four-jointed (Fj) and Approximated (Lawrence et al., 2007; Singh and Mlodzik, 2012; Munoz-Soriano et al., 2012). Fat and Dachsous display opposing

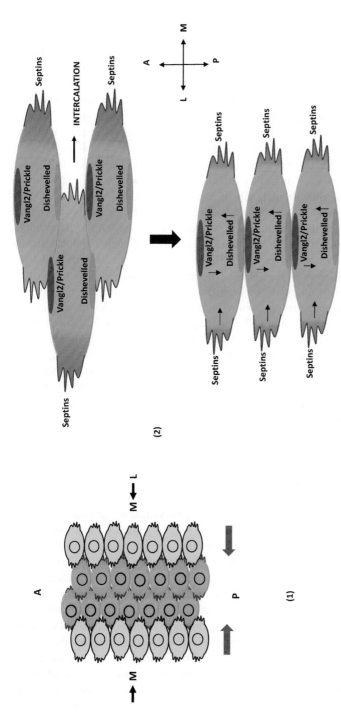

FIGURE 4.3 A schematic representation of the asymmetric planar cell polarity (PCP) protein localization during Convergent extension (CE). (1) Mediolateral intercalation is one of the key processes occurring during CE in which cells intercalate along the mediolateral direction and then redistribute their positions along the AP axis. The red arrow indicates direction of cell movement. A, anterior; L, lateral; M, medial; P, posterior. (2) Three mesodermal cells undergoing mediolateral intercalation are shown. The Vangl2 and Prickle localize to anterior cell membranes, whereas Dvl localizes to posterior cell membrane. The PCP proteins inhibit actomyosin contractility along AP axis by limiting septin to the mediolateral ends of cells.

expression gradients in polarized tissues that are thought to provide directional information. Mice studies show that mice carrying a null mutation in the *Fat4* gene exhibit mild convergent extension defects (Rock et al., 2005). There is only one homolog of *fat* gene identified so far in zebrafish, but its function in PCP is still unexplored (Down et al., 2005). It is established that the Fat/ Dachsous module along with the noncanonical PCP module function together during PCP establishment though their interactions remain complex.

In contrast to *Drosophila* where PCP is exerted across a static epithelial layer, in vertebrates where there is dynamic polarization and migration of cells, whether the observed formation of PCP-protein complexes in apical versus basal regions of migratory cells has remained somewhat unresolved. In dorsal migrating mesodermal cells of *Xenopus laevis*, differential localization of the noncanonical Wnt components, Dvl, Rac, and PKCδ has been observed along the medial and lateral edges, whereas Prickle appears to be localized anteriorly in zebrafish dorsal-migrating mesodermal cells, notochord, and neural tube (Wallingford et al., 2002; Solnica-Krezel, 2005; Komiya and Habas, 2008; Ciruna et al., 2006). The one caveat here is that many of the studies have relied on expressed tagged proteins, which may or may not fully mimic endogenous situations. Notwithstanding these limitations, the analysis of subcellular localization of Dvl remains one of the best studied in vertebrates this far. *Xenopus* Dvl is localized at the plasma membrane and is required to maintain normal lamellipodia stability (Wallingford and Harland, 2001). Overexpression or knockdown of Dvl can lead to disruption of membrane polarization and failure of these lamellipodia to polarize correctly, thereby interfering with ventrolateral convergent extension and further disrupting blastopore closure (Park et al., 2005; Ewald et al., 2004). Thus, the multifunctional protein Dvl plays an important role in coordinating the cell movements and controlling the polarity of the cells, thereby establishing the architecture of the developing embryo. Time-lapse imaging studies have shown that in zebrafish gastrula, Prickle is localized at the anterior edges, whereas the Dvl protein is enriched near the posterior cell edges (Yin et al., 2008). Zebrafish *tribolite* mutant embryos with Vangl2 mutation or Tri/Vangl2 morphants and Knypek mutant embryos with glypican 4 mutations exhibit short body axis and cyclopia due to reduction in convergent extension (Jessen et al., 2002; Heisenberg, 2002; Yin and Solnica-Krezel, 2007). Disruption of Vangl2 function in mouse *looptail* mutants manifests in craniorachischisis, a severe form of neural tube defect characterized by failure in neural tube formation along the entire body axis (Kibar et al., 2001). Similarly, in the mouse node, Prickle2 and Vangl1 are localized at the anterior cell membrane, and Dishevelled2 and Dishevelled3 are localized at the posterior cell membranes (Hashimoto et al., 2010). In zebrafish, Prickle interacts with Dvl, inhibits membrane translocation of Dvl promoting its degradation, suggesting an antagonistic effect to generate PCP (Veeman et al., 2003; Heisenberg and Nusslein-Volhard, 1997). Prickle and Vangl gain of function studies in

Xenopus activates the JNK pathway, which acts downstream of the noncanonical Wnt signaling pathway (Park and Moon, 2002).

ROLE OF CYTOSKELETAL COMPONENTS DURING GASTRULATION

Actin

Actin is one of the major structural components of the cytoskeletal network in all eukaryotic cells and is responsible for a plethora of important cellular processes such as cell motility, cell division, cell signaling, and the establishment and maintenance of cell junctions and cell shape. The inherent ability of cells to migrate is controlled by the actin cytoskeleton which regulates cell polarity, the mechanical force to move the cell, and the organization of adhesions to stabilize the cell during migration (Lecuit and Pilot, 2003). Rearrangement of the actin cytoskeleton during embryogenesis is a major force that drives essential processes including convergent extension and neural tube closure in vertebrates. Therefore, dissecting the dynamics of actin polymerization as well as actin-binding proteins at the subcellular level remains essential to elucidate the mechanism by how noncanonical Wnt signaling regulates vertebrate gastrulation.

During cell migration, two main types of actin structures can be observed on the leading edge of the cell that produces this movement: filopodia and lamellipodia (Albiges-Rizo et al., 2009). Lamellipodia are branched actin filaments that provide force for plasma membrane protrusion during cell migration, which is promoted by the actin-nucleating Arp2/3 complex (Naumanen et al., 2008). During gastrulation in vertebrates, intercalation of mesodermal cells is driven by the PCP pathway. During this process, cells elongate and extend actin-based lamellipodia to interact with laterally adjacent cells such that the cells crawl in between one another (Pfister et al., 2016) (Fig. 4.2). Filopodia are composed of thin membrane protrusions that typically contain bundles of parallel actin filaments, rather than branched, and are used to dynamically extend and retract helping cells to sense their environment and guide migration (Goode and Eck, 2007). Several different types of protein are known to contribute to the formation of filopodia- or lamellipodia-like structures in the cell, such as FMNs, Ena/VASP proteins, and MIM/IRSp53 family proteins that induce direct membrane tubulation (Naumanen et al., 2008). F-actin assembly is compromised in Daam1 deficient cells as these cells display much smaller lamellipodia (Li et al., 2011). Fascin recruits Daam1 to filopodial shafts and silencing of Daam1 also disrupts filopodia formation (Jaiswal et al., 2013).

Another type of cytoskeletal component involved in the regulation of cell movement is the actin stress fiber. Stress fibers are made up of bundles of approximately 10−30 actin filaments with alternating polarity. Stress fibers also contain bipolar arrays of nonmuscle myosin II that alternates between bands of α-actinin that are able to generate a contractile force required for cell movement (Naumanen et al., 2008). There are three different types of stress fibers: ventral, transverse, and dorsal. Ventral stress fibers are anchored at both ends by focal adhesions and are responsible for tail retraction and other contractile movements within the cell. Transverse arcs are curved in nature and cut across dorsal stress fibers, extending back toward the nucleus and upward toward the dorsal surface. The contractile force produced in the transverse arcs is transmitted to the cell-substrate interface through the dorsal stress fibers. These fibers mechanically link to the focal adhesions at one end, and project toward the front of the cell on the other end. In contrast to the other two types of fibers, dorsal fibers do not exhibit colocalization with α-actinin and myosin II. Different types of actin-nucleating proteins are essential for regulating actin fibers in any of these structures. Actin nucleators mainly come in two different subtypes: ARP2/3 Y-branching nucleators and binding partners, and FMNs. Crucial ARP2/3-binding partners include WASP, WAVE, and SCAR proteins (Campellone and Welch, 2010). Unlike Arp2/3 and its interacting proteins, all other known actin nucleators produced unbranched actin filaments. The best-characterized of these nucleators are FMNs, which fall into seven different subclasses based on their FH2 sequences; Dvl-associated activators of morphogenesis (DAAMs), FMNs, Diaphanous (Dia), formin-related proteins in leukocytes (FRLs), delphilin, formin homology domain proteins, and inverted formins (INFs) (Campellone and Welch, 2010).

Microtubules

In addition to the contribution of actin and the microtubule cytoskeleton, the role of adhesion molecules, such as cadherins and integrins, and the importance of membrane dynamics during cell morphogenesis are well-documented (Lecuit and Pilot, 2003). Formation of protrusive structures along the peripheral cell edges of a migrating cell involves a mechanism that orchestrates actin polymerization and membrane deformation. Increasing number of proteins that are able to induce positive membrane curvature, such as proteins belonging to BAR domain family, is known to cause membrane deformation such as Toca-1 (Albiges-Rizo et al., 2009; Suetsugu, 2010). Along with membrane deformation, vesicular transport machinery is also important for transporting signaling molecules to and from the leading edge of the cell via endocytosis and exocytosis, respectively (de Kreuk et al., 2011). Actin filaments, microtubules, and motor proteins are able to generate the force required for vesicular movement (Tsuboi et al., 2009). This vesicular transport machinery has an ability to deform biological membranes into tightly curved vesicles to carry signaling molecules in living cells (Xu et al., 2013). Sequestration of essential components of canonical Wnt signaling, such as glycogen synthase kinase 3, into endosomes is required for proper Wnt signaling (Taelman et al., 2010).

Proteins containing the BAR domain such as syndapin, Toca-1, and sorting nexins are known to deform membranes into tubules (Safari and Suetsugu, 2012). It is proposed that it is the tertiary structure of the BAR protein itself that is able to induce membrane curvature by forming a "banana-shaped" dimer that fits into membrane tubules (Tsuboi et al., 2009; Noguchi, 2016). All BAR domain proteins create concave surfaces at the membrane, but proteins containing Irp53-MIM-homolgy (IMD) domain create a convex surface at the membrane. Indeed, the missing in metastasis protein that contains such an IMD domain has been shown to play a role in anterior neural tube closure in *Xenopus* (Liu et al., 2011).

Planar Cell Polarity Regulation: Cross Talk Between Cytoskeletal Components

FMNs, family of ubiquitous and conserved actin filament assembly factors, have also been found to be implicated in regulation of endocytosis. In cultured cells, both Dia FMNs, mDia1 and mDia2, have been found to be localized to endosomes, suggesting a role of Dia in vesicle trafficking (Goode and Eck, 2007). In addition to playing a critical role in actin polymerization, these mDia class of formins are also involved in regulation of microtubule dynamics. In mammalian fibroblasts, FMNs have been found to play a role in stabilization of microtubules and orientation of the microtubule organizing center and therefore in pathways that direct cell migration (Aspenstrom, 2009). In zebrafish, these Dia proteins controls cell movements and are required for convergent extension during gastrulation (Lai et al., 2008). In nondividing cells, FMNs have been found to regulate the cross talk between microtubules and actin cytoskeleton, which is essential during cell migration. Microtubules provide directional information for motility and help trigger local polymerization leading to membranous protrusions (Goode and Eck, 2007; Bartolini and Gundersen, 2010). However, as compared with their actin nucleating ability, the mechanisms by which FMNs regulate microtubules are far from being fully understood and remains an important area for future investigation.

PLANAR CELL POLARITY IN VERTEBRATES

Based on the localization patterns, PCP proteins could inhibit actin polymerization thereby restricting formation of lamellipodia at the mediolateral cell edges. In *Xenopus*, Dvl-associated activator of morphogenesis, DAAM1, is involved in activation of small GTPases Rho and Rac that promote stress fiber formation and lamellipodia formation respectively, and both small GTPases and DAAM1 are required for convergent extension during gastrulation (Habas et al., 2001). Overexpression and knockdown of Daam1 show severe gastrulation defects in *Xenopus*. Daam regulates processes such as axonal morphogenesis and filopodia formation, and disruption of Daam expression leads to axon growth cone formation defects as well as tracheal defects (Matusek et al., 2008). In neuroepithelial cells, Celsr1 and Dvl recruit the FMN DAAM1 to the AP junction of intercalating cells. Activated DAAM1 in turn interacts with the Rho-GEFS, LFC, p114RhoGEF, and WGEF (Tsuji et al., 2010). This likely activates RhoA and myosin contractility at AP junctions, resulting in convergent extension, apical constriction, and neural plate bending (Nishimura et al., 2012). A similar mechanism was found to drive convergent extension movements of mesenchymal cells during *Xenopus* gastrulation. Septins are cytoskeletal elements that serve as diffusion barriers and resist cell shape deformation (Fung et al., 2014). Knockdown of Sept2 and Sept7 in *Xenopus* resulted in embryos having reduced cell elongation but normal polarization (Kim et al., 2010). In another study, a PCP protein, Fritz interacts with Dvl and helps to recruit septins to mediolateral vertices, where they spatially restrict cortical actomyosin contractility and junctional shrinking to AP cell edges, thus driving cell intercalation (Shindo and Wallingford, 2014). Based on live imaging studies during convergence, epithelial cells in mouse neural plate exhibit tetrads and multicellular rosettes as AP junctions contract and shrink similar to the rosette structures observed in *Drosophila* (Blankenship et al., 2006). Rosette resolution occurs due to formation of new mediolateral junction driving cells to exchange adjacent cells and elongate the tissue. Ptk7 and Vangl2 mutants exhibit open neural tube and shortened body axis due to inhibition of rosette resolution (Williams et al., 2014). These studies show how asymmetric localization of PCP proteins produces collectively polarized cell behaviors via modulation of the cytoskeleton driving convergent extension movements in different cell types. Convergent extension minimizes the distance between the neural folds by lengthening and narrowing the tissues such that allows its apposition along the midline However, convergent extension contributes very little to anterior neural tube closure (Keller et al., 1992), and instead neural fold elevation and neural plate bending are the driving forces (Kee et al., 2008). Noncanonical Wnt component, MIM specifically regulates anterior neural tube closure during gastrulation and not convergent extension (Liu et al., 2011).

CONCLUSIONS

A number of processes including cell migration, gastrulation, and neural tube closure utilize PCP pathway in different cell types and tissues. The concept of PCP applies to understanding of how cells communicate and interact with each other to establish a polarity in the plane of the tissue. In this book chapter, we have highlighted the role of noncanonical Wnt signaling pathway in establishing the PCP in vertebrates. Despite the enormous progress in understanding the Wnt-initiated PCP establishment, little is known about how Wnt morphogens regulate the asymmetrical distribution of these core PCP proteins and the polarized intracellular trafficking of different PCP modulators. Answering these questions will provide us a better understanding of PCP during morphogenesis.

REFERENCES

Albiges-Rizo, C., Destaing, O., Fourcade, B., Planus, E., Block, M.R., 2009. Actin machinery and mechanosensitivity in invadopodia, podosomes and focal adhesions. J. Cell Sci. 122, 3037–3049.

Andersen, M.R., Farooq, M., Koefoed, K., Kjaer, K.W., Simony, A., Christensen, S.T., Larsen, L.A., 2017. Mutation of the planar cell polarity gene VANGL1 in adolescent idiopathic scoliosis. Spine 42, E702–E707.

Andre, P., Wang, Q., Wang, N., Gao, B., Schilit, A., Halford, M.M., Stacker, S.A., Zhang, X., Yang, Y., 2012. The Wnt coreceptor Ryk regulates Wnt/planar cell polarity by modulating the degradation of the core planar cell polarity component Vangl2. J. Biol. Chem. 287, 44518–44525.

Aspenstrom, P., 2009. Formin-binding proteins: modulators of formin-dependent actin polymerization. Biochim. Biophys. Acta 1803, 174–182.

Axelrod, J.D., 2001. Unipolar membrane association of Dishevelled mediates Frizzled planar cell polarity signaling. Genes Dev. 15, 1182–1187.

Bartolini, F., Gundersen, G.G., 2010. Formins and microtubules. Biochim. Biophys. Acta 1803, 164–173.

Beetschen, J.C., 2001. Amphibian gastrulation: history and evolution of a 125 year-old concept. Int. J. Dev. Biol. 45, 771–795.

Blankenship, J.T., Backovic, S.T., Sanny, J.S., Weitz, O., Zallen, J.A., 2006. Multicellular rosette formation links planar cell polarity to tissue morphogenesis. Dev. Cell 11, 459–470.

Campellone, K.G., Welch, M.D., 2010. A nucleator arms race: cellular control of actin assembly. Nat. Rev. Mol. Cell Biol. 11, 237–251.

Chu, C.W., Sokol, S.Y., 2016. Wnt proteins can direct planar cell polarity in vertebrate ectoderm. Elife 5.

Ciruna, B., Jenny, A., Lee, D., Mlodzik, M., Schier, A.F., 2006. Planar cell polarity signalling couples cell division and morphogenesis during neurulation. Nature 439, 220–224.

Davey, C.F., Moens, C.B., 2017. Planar cell polarity in moving cells: think globally, act locally. Development 144, 187–200.

De Felici, M., Siracusa, G., 2000. The rise of embryology in Italy: from the Renaissance to the early 20th century. Int. J. Dev. Biol. 44, 515–521.

de Kreuk, B.J., Nethe, M., Fernandez-Borja, M., Anthony, E.C., Hensbergen, P.J., Deelder, A.M., Plomann, M., Hordijk, P.L., 2011. The F-BAR domain protein PACSIN2 associates with Rac1 and regulates cell spreading and migration. J. Cell Sci. 124, 2375–2388.

Djiane, A., Riou, J., Umbhauer, M., Boucaut, J., Shi, D., 2000. Role of frizzled 7 in the regulation of convergent extension movements during gastrulation in *Xenopus laevis*. Development 127, 3091–3100.

Down, M., Power, M., Smith, S.I., Ralston, K., Spanevello, M., Burns, G.F., Boyd, A.W., 2005. Cloning and expression of the large zebrafish protocadherin gene, Fat. Gene Expr. Patterns 5, 483–490.

Ewald, A.J., Peyrot, S.M., Tyszka, J.M., Fraser, S.E., Wallingford, J.B., 2004. Regional requirements for dishevelled signaling during *Xenopus* gastrulation: separable effects on blastopore closure, mesendoderm internalization and archenteron formation. Development 131, 6195–6209.

Fung, K.Y., Dai, L., Trimble, W.S., 2014. Cell and molecular biology of septins. Int. Rev. Cell Mol. Biol. 310, 289–339.

Glickman, N.S., Kimmel, C.B., Jones, M.A., Adams, R.J., 2003. Shaping the zebrafish notochord. Development 130, 873–887.

Goode, B.L., Eck, M.J., 2007. Mechanism and function of formins in the control of actin assembly. Annu. Rev. Biochem. 76, 593–627.

Gray, R.S., Roszko, I., Solnica-Krezel, L., 2011. Planar cell polarity: coordinating morphogenetic cell behaviors with embryonic polarity. Dev. Cell 21, 120–133.

Habas, R., Kato, Y., He, X., 2001. Wnt/Frizzled activation of Rho regulates vertebrate gastrulation and requires a novel Formin homology protein Daam1. Cell 107, 843–854.

Hale, R., Strutt, D., 2015. Conservation of planar polarity pathway function across the animal kingdom. Annu. Rev. Genet. 49, 529–551.

Hashimoto, M., Shinohara, K., Wang, J., Ikeuchi, S., Yoshiba, S., Meno, C., Nonaka, S., Takada, S., Hatta, K., Wynshaw-Boris, A., Hamada, H., 2010. Planar polarization of node cells determines the rotational axis of node cilia. Nat. Cell Biol. 12, 170–176.

He, X., Semenov, M., Tamai, K., Zeng, X., 2004. LDL receptor-related proteins 5 and 6 in Wnt/beta-catenin signaling: arrows point the way. Development 131, 1663–1677.

Heisenberg, C.P., Nusslein-Volhard, C., 1997. The function of silberblick in the positioning of the eye anlage in the zebrafish embryo. Dev. Biol. 184, 85–94.

Heisenberg, C.P., Tada, M., Rauch, G.J., Saude, L., Concha, M.L., Geisler, R., Stemple, D.L., Smith, J.C., Wilson, S.W., 2000. Silberblick/Wnt11 mediates convergent extension movements during zebrafish gastrulation. Nature 405, 76–81.

Heisenberg, C.P., 2002. Wnt signalling: refocusing on Strabismus. Curr. Biol. 12, R657–R659.

Jaiswal, R., Breitsprecher, D., Collins, A., Correa Jr., I.R., Xu, M.Q., Goode, B.L., 2013. The formin Daam1 and fascin directly collaborate to promote filopodia formation. Curr. Biol. 23, 1373–1379.

Jessen, J.R., Topczewski, J., Bingham, S., Sepich, D.S., Marlow, F., Chandrasekhar, A., Solnica-Krezel, L., 2002. Zebrafish trilobite identifies new roles for Strabismus in gastrulation and neuronal movements. Nat. Cell Biol. 4, 610–615.

Kee, N., Wilson, N., De Vries, M., Bradford, D., Key, B., Cooper, H.M., 2008. Neogenin and RGMa control neural tube closure and neuroepithelial morphology by regulating cell polarity. J. Neurosci. 28, 12643–12653.

Keller, R., Shih, J., Sater, A., 1992. The cellular basis of the convergence and extension of the Xenopus neural plate. Dev. Dyn. 193, 199–217.

Keller, R., Davidson, L.A., Shook, D.R., 2003. How we are shaped: the biomechanics of gastrulation. Differentiation 71, 171–205.

Kibar, Z., Vogan, K.J., Groulx, N., Justice, M.J., Underhill, D.A., Gros, P., 2001. Ltap, a mammalian homolog of Drosophila Strabismus/Van Gogh, is altered in the mouse neural tube mutant Loop-tail. Nat. Genet. 28, 251–255.

Kilian, B., Mansukoski, H., Barbosa, F.C., Ulrich, F., Tada, M., Heisenberg, C.P., 2003. The role of Ppt/Wnt5 in regulating cell shape and movement during zebrafish gastrulation. Mech. Dev. 120, 467–476.

Kim, S.K., Shindo, A., Park, T.J., Oh, E.C., Ghosh, S., Gray, R.S., Lewis, R.A., Johnson, C.A., Attie-Bittach, T., Katsanis, N., Wallingford, J.B., 2010. Planar cell polarity acts through septins to control collective cell movement and ciliogenesis. Science 329, 1337–1340.

Komiya, Y., Habas, R., 2008. Wnt signal transduction pathways. Organogenesis 4, 68–75.

Lai, S.L., Chan, T.H., Lin, M.J., Huang, W.P., Lou, S.W., Lee, S.J., 2008. Diaphanous-related formin 2 and profilin I are required for gastrulation cell movements. PLoS One 3, e3439.

Lawrence, P.A., Shelton, P.M., 1975. The determination of polarity in the developing insect retina. J. Embryol. Exp. Morphol. 33, 471–486.

Lawrence, P.A., Casal, J., Struhl, G., 2002. Towards a model of the organisation of planar polarity and pattern in the *Drosophila* abdomen. Development 129, 2749–2760.

Lawrence, P.A., Struhl, G., Casal, J., 2007. Planar cell polarity: one or two pathways? Nat. Rev. Genet. 8, 555–563.

Lecuit, T., Pilot, F., 2003. Developmental control of cell morphogenesis: a focus on membrane growth. Nat. Cell Biol. 5, 103–108.

Lee, J.Y., Harland, R.M., 2007. Actomyosin contractility and microtubules drive apical constriction in *Xenopus* bottle cells. Dev. Biol. 311, 40–52.

Li, L., Yuan, H., Xie, W., Mao, J., Caruso, A.M., Mcmahon, A., Sussman, D.J., Wu, D., 1999. Dishevelled proteins lead to two signaling pathways. Regulation of LEF-1 and c-Jun N-terminal kinase in mammalian cells. J. Biol. Chem. 274, 129–134.

Li, D., Hallett, M.A., Zhu, W., Rubart, M., Liu, Y., Yang, Z., Chen, H., Haneline, L.S., Chan, R.J., Schwartz, R.J., Field, L.J., Atkinson, S.J., Shou, W., 2011. Dishevelled-associated activator of morphogenesis 1 (Daam1) is required for heart morphogenesis. Development 138, 303–315.

Liu, W., Komiya, Y., Mezzacappa, C., Khadka, D.K., Runnels, L., Habas, R., 2011. MIM regulates vertebrate neural tube closure. Development 138, 2035–2047.

Lu, W., Yamamoto, V., Ortega, B., Baltimore, D., 2004a. Mammalian Ryk is a Wnt coreceptor required for stimulation of neurite outgrowth. Cell 119, 97–108.

Lu, X., Borchers, A.G., Jolicoeur, C., Rayburn, H., Baker, J.C., Tessier-Lavigne, M., 2004b. PTK7/CCK-4 is a novel regulator of planar cell polarity in vertebrates. Nature 430, 93–98.

Marlow, F., Topczewski, J., Sepich, D., Solnica-Krezel, L., 2002. Zebrafish Rho kinase 2 acts downstream of Wnt11 to mediate cell polarity and effective convergence and extension movements. Curr. Biol. 12, 876–884.

Matusek, T., Gombos, R., Szecsenyi, A., Sanchez-Soriano, N., Czibula, A., Pataki, C., Gedai, A., Prokop, A., Rasko, I., Mihaly, J., 2008. Formin proteins of the DAAM subfamily play a role during axon growth. J. Neurosci. 28, 13310–13319.

Mlodzik, M., 2002. Planar cell polarization: do the same mechanisms regulate *Drosophila* tissue polarity and vertebrate gastrulation? Trends Genet. 18, 564–571.

Munoz-Soriano, V., Belacortu, Y., Paricio, N., 2012. Planar cell polarity signaling in collective cell movements during morphogenesis and disease. Curr. Genom. 13, 609–622.

Naumanen, P., Lappalainen, P., Hotulainen, P., 2008. Mechanisms of actin stress fibre assembly. J. Microsc. 231, 446–454.

Nishimura, T., Honda, H., Takeichi, M., 2012. Planar cell polarity links axes of spatial dynamics in neural-tube closure. Cell 149, 1084–1097.

Nishita, M., Yoo, S.K., Nomachi, A., Kani, S., Sougawa, N., Ohta, Y., Takada, S., Kikuchi, A., Minami, Y., 2006. Filopodia formation mediated by receptor tyrosine kinase Ror2 is required for Wnt5a-induced cell migration. J. Cell Biol. 175, 555–562.

Noguchi, H., 2016. Membrane tubule formation by banana-shaped proteins with or without transient network structure. Sci. Rep. 6, 20935.

Park, M., Moon, R.T., 2002. The planar cell-polarity gene stbm regulates cell behaviour and cell fate in vertebrate embryos. Nat. Cell Biol. 4, 20–25.

Park, T.J., Gray, R.S., Sato, A., Habas, R., Wallingford, J.B., 2005. Subcellular localization and signaling properties of dishevelled in developing vertebrate embryos. Curr. Biol. 15, 1039–1044.

Park, T.J., Haigo, S.L., Wallingford, J.B., 2006. Ciliogenesis defects in embryos lacking inturned or fuzzy function are associated with failure of planar cell polarity and Hedgehog signaling. Nat. Genet. 38, 303−311.

Pfister, K., Shook, D.R., Chang, C., Keller, R., Skoglund, P., 2016. Molecular model for force production and transmission during vertebrate gastrulation. Development 143, 715−727.

Rock, R., Schrauth, S., Gessler, M., 2005. Expression of mouse dchs1, fjx1, and fat-j suggests conservation of the planar cell polarity pathway identified in Drosophila. Dev. Dyn. 234, 747−755.

Rolo, A., Skoglund, P., Keller, R., 2009. Morphogenetic movements driving neural tube closure in Xenopus require myosin IIB. Dev. Biol. 327, 327−338.

Safari, F., Suetsugu, S., 2012. The BAR domain superfamily proteins from subcellular structures to human diseases. Membr. (Basel) 2, 91−117.

Sasai, N., Nakazawa, Y., Haraguchi, T., Sasai, Y., 2004. The neurotrophin-receptor-related protein NRH1 is essential for convergent extension movements. Nat. Cell Biol. 6, 741−748.

Seifert, J.R., Mlodzik, M., 2007. Frizzled/PCP signalling: a conserved mechanism regulating cell polarity and directed motility. Nat. Rev. Genet. 8, 126−138.

Shindo, A., Wallingford, J.B., 2014. PCP and septins compartmentalize cortical actomyosin to direct collective cell movement. Science 343, 649−652.

Singh, J., Mlodzik, M., 2012. Planar cell polarity signaling: coordination of cellular orientation across tissues. Wiley Interdiscip. Rev. Dev. Biol. 1, 479−499.

Solnica-Krezel, L., Sepich, D.S., 2012. Gastrulation: making and shaping germ layers. Annu. Rev. Cell Dev. Biol. 28, 687−717.

Solnica-Krezel, L., 2005. Conserved patterns of cell movements during vertebrate gastrulation. Curr. Biol. 15, R213−R228.

Strutt, D.I., 2001. Asymmetric localization of frizzled and the establishment of cell polarity in the Drosophila wing. Mol. Cell 7, 367−375.

Suetsugu, S., 2010. The proposed functions of membrane curvatures mediated by the BAR domain superfamily proteins. J. Biochem. 148, 1−12.

Taelman, V.F., Dobrowolski, R., Plouhinec, J.L., Fuentealba, L.C., Vorwald, P.P., Gumper, I., Sabatini, D.D., De Robertis, E.M., 2010. Wnt signaling requires sequestration of glycogen synthase kinase 3 inside multivesicular endosomes. Cell 143, 1136−1148.

Tahinci, E., Thorne, C.A., Franklin, J.L., Salic, A., Christian, K.M., Lee, L.A., Coffey, R.J., Lee, E., 2007. Lrp6 is required for convergent extension during Xenopus gastrulation. Development 134, 4095−4106.

Taylor, J., Abramova, N., Charlton, J., Adler, P.N., 1998. Van Gogh: a new Drosophila tissue polarity gene. Genetics 150, 199−210.

Theisen, H., Purcell, J., Bennett, M., Kansagara, D., Syed, A., Marsh, J.L., 1994. Dishevelled is required during wingless signaling to establish both cell polarity and cell identity. Development 120, 347−360.

Tree, D.R., Ma, D., Axelrod, J.D., 2002a. A three-tiered mechanism for regulation of planar cell polarity. Semin. Cell Dev. Biol. 13, 217−224.

Tree, D.R., Shulman, J.M., Rousset, R., Scott, M.P., Gubb, D., Axelrod, J.D., 2002b. Prickle mediates feedback amplification to generate asymmetric planar cell polarity signaling. Cell 109, 371−381.

Tsuboi, S., Takada, H., Hara, T., Mochizuki, N., Funyu, T., Saitoh, H., Terayama, Y., Yamaya, K., Ohyama, C., Nonoyama, S., Ochs, H.D., 2009. FBP17 mediates a common molecular step in the formation of podosomes and phagocytic cups in macrophages. J. Biol. Chem. 284, 8548−8556.

Tsuji, T., Ohta, Y., Kanno, Y., Hirose, K., Ohashi, K., Mizuno, K., 2010. Involvement of p114-RhoGEF and Lfc in Wnt-3a- and dishevelled-induced RhoA activation and neurite retraction in N1E-115 mouse neuroblastoma cells. Mol. Biol. Cell 21, 3590—3600.

Usui, T., Shima, Y., Shimada, Y., Hirano, S., Burgess, R.W., Schwarz, T.L., Takeichi, M., Uemura, T., 1999. Flamingo, a seven-pass transmembrane cadherin, regulates planar cell polarity under the control of Frizzled. Cell 98, 585—595.

Veeman, M.T., Slusarski, D.C., Kaykas, A., Louie, S.H., Moon, R.T., 2003. Zebrafish prickle, a modulator of noncanonical Wnt/Fz signaling, regulates gastrulation movements. Curr. Biol. 13, 680—685.

Vinson, C.R., Adler, P.N., 1987. Directional non-cell autonomy and the transmission of polarity information by the frizzled gene of *Drosophila*. Nature 329, 549—551.

Wallingford, J.B., Harland, R.M., 2001. *Xenopus* Dishevelled signaling regulates both neural and mesodermal convergent extension: parallel forces elongating the body axis. Development 128, 2581—2592.

Wallingford, J.B., Harland, R.M., 2002. Neural tube closure requires Dishevelled-dependent convergent extension of the midline. Development 129, 5815—5825.

Wallingford, J.B., Fraser, S.E., Harland, R.M., 2002. Convergent extension: the molecular control of polarized cell movement during embryonic development. Dev. Cell 2, 695—706.

Wang, J., Hamblet, N.S., Mark, S., Dickinson, M.E., Brinkman, B.C., Segil, N., Fraser, S.E., Chen, P., Wallingford, J.B., Wynshaw-Boris, A., 2006. Dishevelled genes mediate a conserved mammalian PCP pathway to regulate convergent extension during neurulation. Development 133, 1767—1778.

Wen, S., Zhu, H., Lu, W., Mitchell, L.E., Shaw, G.M., Lammer, E.J., Finnell, R.H., 2010. Planar cell polarity pathway genes and risk for spina bifida. Am. J. Med. Genet. A 152A, 299—304.

Williams, M., Yen, W., Lu, X., Sutherland, A., 2014. Distinct apical and basolateral mechanisms drive planar cell polarity-dependent convergent extension of the mouse neural plate. Dev. Cell 29, 34—46.

Witt, E., 2008. Form—a matter of generation: the relation of generation, form, and function in the epigenetic theory of Caspar F. Wolff. Sci. Context 21, 649—664.

Wu, J., Roman, A.C., Carvajal-Gonzalez, J.M., Mlodzik, M., 2013. Wg and Wnt4 provide long-range directional input to planar cell polarity orientation in *Drosophila*. Nat. Cell Biol. 15, 1045—1055.

Xu, P., Baldridge, R.D., Chi, R.J., Burd, C.G., Graham, T.R., 2013. Phosphatidylserine flipping enhances membrane curvature and negative charge required for vesicular transport. J. Cell Biol. 202, 875—886.

Yin, C., Solnica-Krezel, L., 2007. Convergence and extension movements affect dynamic notochord-somite interactions essential for zebrafish slow muscle morphogenesis. Dev. Dyn. 236, 2742—2756.

Yin, C., Kiskowski, M., Pouille, P.A., Farge, E., Solnica-Krezel, L., 2008. Cooperation of polarized cell intercalations drives convergence and extension of presomitic mesoderm during zebrafish gastrulation. J. Cell Biol. 180, 221—232.

Chapter 5

Polarized Membrane Trafficking in Development and Disease: From Epithelia Polarization to Cancer Cell Invasion

Erik Linklater, Cayla E. Jewett, Rytis Prekeris
University of Colorado Anschutz Medical Campus, Aurora, CO, United States

INTRODUCTION

Generation and maintenance of cell polarity is one of the most fundamental properties of multicellular organisms and failure of any aspect of cell polarization can lead to severe consequences that affect the function of individual cells, tissues, or the entire organism. In some cases cells terminally differentiate to generate permanently polarized cells with specific functions, such as neurons or epithelial cells. Additionally, most cells are also capable of producing transient polarized structures, such as the ingression furrow during cell division or leading edge or motile protrusions during cell migration. Interestingly, despite big differences in the initiation, formation, and regulation of permanent and transient cell polarity, the core machinery mediating cell polarization appears to have many common features. Studies from laboratories over the last few decades started identifying the mechanisms that govern the formation of cell polarity. While cell polarization is a very complex cellular event, it has become clear that one of the key players of this process is polarized membrane and protein transport. It is now well established that there are two pathways that lead to polarized delivery of proteins; direct transport of newly synthesized proteins from the trans-Golgi network (TGN) or polarized resorting and redelivery of proteins via domain-specific endocytic transport pathways. In both cases, this polarized protein and membrane transport relies on the generation of cargo-specific transport vesicles and targeting these vesicles to a specific subcellular domain. It is also becoming clear that

Cell Polarity in Development and Disease. https://doi.org/10.1016/B978-0-12-802438-6.00005-X

polarized membrane traffic is closely integrated with localized microtubule and actin cytoskeleton dynamics. On one hand, microtubules and actin microfilaments direct the transport of endosome or TGN-derived transport vesicles to their final destinations. On the other hand, it has been now demonstrated that these transport vesicles often carry various cytoskeleton modulators that directly affect localized filamentous actin and microtubule growth and depolymerization. Thus, regulation and dynamics of this membrane transport and cytoskeleton interplay has become one of the major research topics in the cell polarity field. The size limitations of this review do not permit us to systematically describe all the new data regarding membrane transport and cell polarity. Instead, we will focus on the role of membrane transport and cytoskeleton cross talk only in two emerging cell polarity fields. In the first half of this review, we will describe the role and regulation of apical endocytic transport during epithelia polarization. In the second half of the review, we will describe the role of cross talk between the actin cytoskeleton and membrane transport during formation and extension of invadopodia, transient but highly polarized structures formed during cell migration.

EPITHELIAL POLARITY AND LUMEN FORMATION

Epithelial cells are one of the major polarized cell types and are key to the function of most tissues and organs. Consequently, much effort has been dedicated to understanding the machinery that is required and mediates the polarization of individual epithelial cells. Importantly, during tissue morphogenesis, epithelial cells need to coordinate their polarization to form intricate systems of tubules and end buds. While some of these epithelial structures assume their final form during development, many other epithelial tubules, such as hepatic canaliculi or mammary ducts, can be constantly remodeled even in adult organisms (Sternlicht et al., 2006). Thus, it is now clear that in addition to machinery mediating polarization of individual cells, there needs to be integration of these polarization events across multiple cells to lead to proper formation of epithelial tissues. Here we will describe the latest findings about the involvement of polarized membrane transport and the cytoskeleton during polarization of individual cells as well as during formation of multicellular epithelial structures.

Membrane Transport and Polarization of Individual Epithelial Cells

Individual epithelial cells are polarized to maintain the identity and distinct compositions of the apical and basolateral sides of the cell. This apicobasal polarization allows for a single cell to adapt to two very different environments. The apical side faces the lumen and therefore must respond to gaseous or aqueous mediums, whereas the basolateral domain is surrounded by neighboring cells or an extracellular matrix (ECM). The apical and basolateral domains are separated by a group of scaffolding proteins that form a structure called the tight junction (TJ) (Zihni et al., 2016). TJs act as a diffusion barrier to prevent mixing of apical and basolateral membrane components and also function as an intercellular seal. TJs are composed of two transmembrane protein families, claudins and occludins, which form a dimer between two neighboring cells in the intercellular space. A peripheral membrane protein, zonula occludens (ZO-1, ZO-2, ZO-3) can bind both claudins/occludins and the actin cytoskeleton, thus providing a link between actin and the neighboring cells.

Since the formation of apicobasal polarity and TJ-based separation of apical and basolateral domains are the key to the function of epithelial cells, there are correspondingly multiple and complex molecular pathways that ensure the fidelity of epithelial cell polarization. There are three highly conserved key polarity complexes that regulate this process and are responsible for giving the plasma membrane its apical or basolateral identity. The first is the apically localized Crumbs complex composed of Crumbs/Pals1/PatJ (Assemat et al., 2008). The second protein complex, the basolateral Scribble complex, contains Scribble/Dlg/Lgl (Assemat et al., 2008). Finally, the Par complex is composed of Par3/Par6/aPKC/Cdc42 and localizes laterally to TJs (Assemat et al., 2008). These three polarity complexes are essential for cell polarization and misexpression or inactivation of any of these complexes results in reorientation or complete loss of apicobasal polarity. In the last few years, several excellent reviews have been written regarding the function of these polarity complexes (Assemat et al., 2008; Tepass, 2012; Roman-Fernandez and Bryant, 2016), thus this review instead will focus on the role of membrane transport during epithelia polarization.

Because polarized cells are physically separated into distinct domains, there is a need for selective delivery of membrane and proteins to either the apical or basolateral domain. This is accomplished by polarized membrane trafficking via domain-specific organelles, such as apical and basolateral endosomes, which are regulated by Rab GTPases (Apodaca et al., 2012). Rab proteins are a family of small monomeric GTPases that are master regulators of membrane transport. Rabs recruit proteins called effectors that together give a vesicle a "zip code" that targets it to a specific location within the cell. Although there are close to 70 different Rab proteins in mammalian cells, only a dozen or so have been implicated in regulating apicobasal polarity. Through a functional screen of 60 different Rab proteins, Rab3, Rab4, Rab5, Rab8a, Rab11a, Rab12, Rab13, Rab14, Rab15, Rab17, Rab19, Rab25, Rab27a, Rab32, and Rab35 were shown to disrupt glycoprotein-135 (GP135 also called podocalyxin) trafficking to the apical surface in MDCK II cells (Mrozowska and Fukuda, 2016). While there is currently little known about many of these Rabs, a few have been identified as key regulators of epithelial polarity.

In particular, Rab11 has emerged as a central regulator of apical protein transport in vitro and in vivo (Jing and Prekeris, 2009; Sobajima et al., 2014; Alvers et al., 2014). Most vertebrate epithelial cells express three Rab11 subfamily members, Rab11a, Rab11b, and Rab25 (sometimes referred to as Rab11c) (Casanova et al., 1999; Lapierre et al., 2003; Chavrier et al., 1990; Goldenring et al., 1993). While the involvement of Rab25 in polarized epithelial transport remains to be understood, both Rab11a and Rab11b play important (and partially redundant) roles in mediating protein targeting to apical recycling endosomes and apical plasma membrane (Goldenring et al., 1993; Junutula et al., 2004). Rab11a/b binds to Rab11 Family Interacting Proteins (FIPs), which recruit additional polarized trafficking machinery and mediate multiple apical targeting steps (Prekeris, 2003; Baetz and Goldenring, 2013; Hales et al., 2001). Additionally, Rab11 and its effector FIP2, Rab8, and Rab10 can also bind to the actin motor Myosin-5b (Roland et al., 2011; Lapierre et al., 2001; Hales et al., 2002). Interestingly, Myosin-5b is also required for transporting and/or tethering apical transport vesicles to the actin-rich apical plasma membrane (Roland et al., 2011) with mutations in Myosin-5b leading to microvillus inclusion disease, a congenital disorder that affects apical transport and microvilli formation in enterocytes (Erickson et al., 2008; Muller et al., 2008; Knowles et al., 2014). In addition to FIPs, Rab11a/b also binds to Sec15, a component of the Exocyst complex (Zhang et al., 2004), which is a well-established tethering factor that mediates polarized transport to plasma membrane (Hsu et al., 2004).

Rab GTPases are often promiscuous proteins able to bind to multiple effectors and are activated by multiple guanine nucleotide exchange factors (GEFs). In addition, it is becoming increasingly clear that membrane trafficking events may occur through "Rab cascades," whereby one Rab protein will recruit an effector, which is a GEF activating a second Rab protein, which in turn recruits another effector, and so on. These Rab cascades were first discovered through Rab5 conversion to Rab7 (Rink et al., 2005), but several other pathways have since been elucidated (Novick, 2016; Hutagalung and Novick, 2011). Rab11 has been implicated in mediating one Rab cascade that is involved in apical primary cilia formation and may also have additional roles in apical targeting. During apical primary cilia formation, Rab11 binds to Rabin8, a GEF for Rab8, another player in polarized vesicle transport (Knodler et al., 2010). It was proposed that this Rab11-dependent recruitment of Rabin8 to apical endosomes then leads to activation and recruitment of Rab8, which in turn regulates protein targeting during ciliogenesis (Knodler et al., 2010). Interestingly, Rab11 was also implicated in a "Rab-Arf cascade" whereby Rab11 and its effector protein FIP3 recruit Arf4, another small monomeric GTPase that regulates protein targeting to sensory cilia (Mazelova et al., 2009; Wang and Deretic, 2015) Thus, it is becoming clear that membrane transport involves an intricate network of Rab and Arf GTPases and their effectors for specific targeting to a certain site in polarized cells.

Polarized Membrane Transport and Apical Lumen Formation During Epithelial Tissue Morphogenesis

During epithelial tissue morphogenesis and remodeling, epithelial cell polarization is a highly coordinated event that leads to formation of an apical lumen. The establishment of an apical lumen is a critical step during development resulting in the formation of all hollow organs. How single lumens arise from nonpolarized cells, called de novo lumen formation, have been a topic under recent investigation. One well-accepted model is the midbody-dependent hollowing model (Datta et al., 2011; Li et al., 2014). This model consists of three steps (Fig. 5.1). First, there is a symmetry-breaking event in which a single nonpolarized cell divides, and the midbody from that cell division provides a landmark for the establishment of apical membrane identity. During the second step, apical endosomes are targeted to the midbody and fuse with the plasma membrane domain, known as the apical membrane initiation site or AMIS. This delivery of cargo and fusion of vesicles with the cell membrane results in formation of a nascent lumen. Finally, subsequent cell divisions expand the luminal space resulting in the formation of a polarized multicellular epithelial structure (Fig. 5.1) (Li et al., 2014; Datta et al., 2011).

To ensure the fidelity of lumenogenesis, there is a dynamic interplay between polarity proteins, trafficking proteins, and the cytoskeletal network. The Rab11 GTPase and its effector FIP5 were recently shown to mediate apical transport to the AMIS during lumenogenesis in vitro (Willenborg et al., 2011) (Fig. 5.2). The Rab11/FIP5 complex sequentially interacts with sorting nexin 18 (SNX18), kinesin-2, and cingulin to mediate apical vesicle formation, transport, and targeting to AMIS (Fig. 5.2) (Willenborg et al., 2011; Li et al., 2014). This process is also regulated by Myosin-5b, presumably through binding to Rab11 and possibly Rab8 (Wakabayashi et al., 2005; Li et al., 2007; Vogel et al., 2015).

As in many other polarized cells, the actin and microtubule cytoskeleton closely interacts with and regulates Rab-dependent apical membrane transport. Recently we have shown that the Rab11/FIP5 vesicle tether cingulin directly binds to central spindle microtubules, presumably by interacting with glutamylated tubulin C-terminal tails (Fig. 5.2) (Mangan et al., 2016). Once recruited to the AMIS (by binding to ZO-1 and tubulin), cingulin activates Arp2/3-dependent branched actin polymerization, which in turn recruits more ZO-1 (and consequently cingulin) as well as mediates Myosin-5b-dependent apical vesicle transport/tethering (Mangan et al., 2016). This type of positive-feedback loop enhances AMIS formation and apical vesicle targeting.

The key to midbody-dependent lumen formation is the spatiotemporal regulation of apical vesicle transport. During cell division, apical cargo transport needs to be delayed until late telophase to allow formation of the midbody-associated AMIS. At least in part this delay in apical transport is regulated by FIP5 phosphorylation. We have recently shown that during

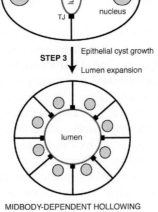

MIDBODY-DEPENDENT HOLLOWING

FIGURE 5.1 Midbody-dependent hollowing model of lumen formation. A single non-polarized cell divides and forms a midbody, which functions as a spatial cue for creation of the apical site (Step 1). Apical endosomes are then transported to the midbody, establishing the apical membrane initiation site (AMIS) (Step 2a). Consequently, the delivery and fusion of apical endosomes results in nascent lumen formation (Step 2b). Subsequent cell divisions expand the luminal space and increase the size of the cyst (Step 3).

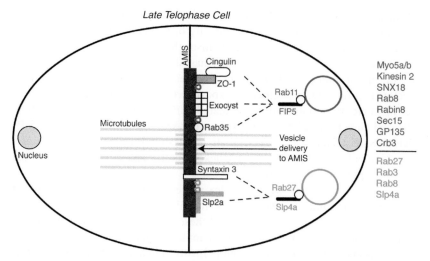

FIGURE 5.2 **Molecular mechanisms behind apical lumen formation.** During late telophase, Rab11/FIP5 endosomes and their binding partners (proteins listed in red) are transported to the apical membrane initiation site (AMIS) where they are tethered by several complexes, including ZO-1/Cingulin, the Exocyst, and Rab35, mediating vesicle fusion with the AMIS. Additionally, Rab27/Slp4a endosomes and their binding partners (listed in green) are also targeted to the AMIS through tethering with Slp2a and Syntaxin-3. Dotted lines show tethering targets. Proteins that localize to the AMIS are in blue.

metaphase and anaphase FIP5 is phosphorylated at T276 by glycogen synthase kinase-3, which keeps Rab11/FIP5 endosomes localized to the centrosomes by inhibiting FIP5 binding to SNX18. During late telophase, FIP5 is dephosphorylated allowing it to interact with SNX18 to promote budding and transport of apical endosomes to the newly formed AMIS (Li et al., 2014). Interestingly, inhibition of Rab11-dependent endocytic recycling during metaphase and anaphase was also observed in nonepithelial cells (Das et al., 2014; Hehnly and Doxsey, 2014; Wilson et al., 2005; Schiel et al., 2013), suggesting that the regulation of endocytic recycling during cell division may have other functions in addition to regulating epithelial targeting.

Rab35 is another Rab that is required for protein targeting during apical lumen formation (Mrozowska and Fukuda, 2016; Klinkert et al., 2016) (Fig. 5.2). Similar to Rab11, Rab35 was shown to mediate apical cargo transport to the midbody-associated AMIS (Klinkert et al., 2016). However, Rab35 localizes to the AMIS rather than apical transport vesicles. Thus, it was proposed that Rab35 directly tethers apical transport vesicles by either directly binding to GP135 (known apical vesicle cargo protein) or by interacting with ACAP2 (Fig. 5.2) (Klinkert et al., 2016; Mrozowska and Fukuda, 2016).

Several other important players in apical lumen formation have recently come to light. Rab14 was found to act upstream of trafficking events in the regulation of membrane lipid composition at the apical surface, by modulating

Cdc42 activation and/or midbody positioning (Lu and Wilson, 2016). However, it is unclear whether it does so directly or indirectly, and further study is needed to determine the precise mechanism through which Rab14 acts. The synaptotagmin-like protein (Slp) family has also emerged as putative regulators of apical transport during lumenogenesis. Slps are a group of Rab effectors that function as a tether for exocytic vesicles. It was shown that Slp2a and Slp4a are transcriptionally upregulated during lumen formation in MDCK cells and direct Rab27 vesicles to apical membrane enriched in PI(4,5)P$_2$ (Galvez-Santisteban et al., 2012). Slp4a is a Rab27/Rab8/Rab3b effector and must bind all three Rabs to be trafficked to the apical surface (Fig. 5.2) (Galvez-Santisteban et al., 2012).

Apical lumen formation is a complex process that requires the coordinated regulation and timing of the polarity machinery, trafficking proteins, and cytoskeleton. Functional loss of many of these proteins results in disruption of the apical membrane and generation of multilumenal cysts. The number of different Rab proteins and their effectors involved in apical–basal polarity and lumen formation is overwhelming and raises the question of why such an intricate process is necessary. Are each of these Rab-labeled vesicles carrying different cargo, or is there a coincidence detection system to ensure fidelity of trafficking and formation of a single lumen? Furthermore, in addition to Rab11 and Rab35 there are a number of other less understood Rabs that appear to perturb delivery of GP135 to the apical membrane in 3D MDCK cysts (Mrozowska and Fukuda, 2016). Since there are no published studies that dissect how these Rabs function, it remains unclear whether they directly affect protein targeting during lumenogenesis and how they relate to Rab11 and Rab35-dependent apical transport.

Regulation of De Novo Lumen Formation In Vivo

Most work in understanding de novo apical lumen formation comes from 3D tissue culture systems. While this model is widely accepted and provides valuable mechanistic information, it will be important to further validate this work in model organisms where multiple organ systems function in synchrony and compensatory mechanisms may be at work. One of the simplest in vivo systems to study lumenogenesis is in the *Caenorhabditis elegans* excretory canal formation. The canal is comprised entirely of a single cell that sends out hollow tubules to create an H-shape extending the length of the worm. These tubes are epithelial in character, with the apical Crumbs and PAR complexes at the luminal side (Bossinger et al., 2001; Mancuso et al., 2012; Armenti et al., 2014). Lumen formation begins when specialized vesicles at the apical surface tether and fuse, aided by the Exocyst complex. The asymmetric localization of the Exocyst complex is defined by PAR proteins. However, it is still unclear how the apical domain is initially established and much detail is lacking about the vesicles that deliver apical membrane.

The *Drosophila* tracheal system is probably the best characterized model for tube formation. This organ system spans the entire larvae and provides oxygen to every cell. Tracheal tubes can be multicellular or unicellular and utilize several different mechanisms to form lumens. The terminal cells, which are located at the ends of the tracheal branches and deliver oxygen to individual cells, specifically undergo de novo lumen formation. Similar to 3D tissue culture models, Rab trafficking, the Exocyst, and PAR proteins are necessary for lumen formation in terminal cells. A Rab35 GAP, Whacked, localizes to the apical membrane, and this membrane trafficking pathway is responsible for controlling where and how much a tube grows (Schottenfeld-Roames and Ghabrial, 2012). The Exocyst is required for vesicle tethering and fusion during lumen formation and is facilitated by the PAR polarity complex (Jones and Metzstein, 2011; Jones et al., 2014). In addition, either the Rab11 recycling endosome-derived trafficking pathway or the Rab10 Golgi-derived pathway are necessary for terminal cell outgrowth, suggesting that unlike tissue culture models, the two vesicle trafficking pathways are redundant (Jones et al., 2014).

A third model system that is becoming more widely used to study lumen formation is the intestinal tract of zebrafish, *Danio rerio*. The intestinal system starts out as a solid cord of endodermal cells that have migrated to the midline. Around 2 days postfertilization (dpf), these endodermal cells develop an apical surface and immediately begin to form small lumens. These cells are also highly proliferative. At 3 dpf, the small lumens have coalesced to form a single large lumen and cells continue to proliferate and expand the lumen. The endoderm becomes epithelial in character, adopting a columnar shape and establishing strong apical—basal polarity. By 4 dpf, the intestinal tract is a completely open tube from mouth to anus, and specialized compartments have developed. The entire development of the intestinal tract occurs in the rostral to caudal direction (Ng et al., 2005).

In zebrafish, it is still unclear how lumen formation occurs at the molecular level. Evidence supporting de novo establishment of an apical domain (Horne-Badovinac et al., 2001; Ng et al., 2005), and lack of apoptosis in lumen formation (Ng et al., 2005) suggest that expansion of the apical membrane may result from molecular factors found in tissue culture and other in vivo models, such as the Exocyst complex, Rab proteins, and the cytoskeleton. Indeed, it has been shown that the apical polarity protein aPKC is necessary for single lumen formation in the zebrafish intestine (Horne-Badovinac et al., 2001). Further research is needed to determine the precise molecular mechanisms.

One major difference between in vitro cultured lumens and many in vivo models is that during epithelial tissue morphogenesis there are at least two distinct steps of lumenogenesis, formation of nascent mini lumens and coalescence of these mini lumens into the final luminal structure. While the molecular machinery governing mini lumen formation is beginning to emerge, what mediates lumen coalescence is unclear largely because there is no comparable in vitro experimental model to study the process of coalescence. Thus, development of new 3D tissue culture models may help elucidate these processes.

POLARIZED MEMBRANE TRANSPORT DURING CANCER METASTASIS AND DEVELOPMENT

Cell migration in vivo requires the generation of polarized actin-rich protrusions that are needed to both remodel the ECM and to sense extracellular guidance cues. It is becoming clear that the formation and function of these migratory protrusions rely on localized cytoskeleton remodeling as well as polarized membrane transport of intracellular cargo. In this section we will summarize the newest findings defining molecular machinery governing polarized membrane transport to these migratory protrusions and the role of this transport in regulating cell migration during development and carcinogenesis.

Cells make many different migratory protrusions that interact with the extracellular environment during cell migration through the ECM. Two related types of such protrusions are called podosomes and invadopodia. They were initially described during the mid-1980s when several groups studying Src-transformed fibroblasts reported the presence of actin-rich ventral protrusions that were feetlike in appearance and termed the structures podosomes (Tarone et al., 1985). On subsequent discovery that these structures also occurred at areas of ECM degradation, they were called invadopodia (Chen et al., 1985; Chen, 1989). Similar actin-rich protrusions have been identified in other cells, including both multiple types of normal cells as well as cancer cells, and the term podosome is now used to refer to these structures in normal cells, while invadopodia refer to such structures in cancer cells. The general features of these two structures share many commonalties that highlight the shared nature of their form and function yet also differ in several key aspects that ultimately distinguish them as distinct entities.

Podosomes and Cell Migration

The last several decades have defined a role for podosomes in many types of cells, including cells of myeloid lineage such as dendritic cells and osteoclasts, as well as smooth muscle cells (Kaverina et al., 2003) and endothelial cells (Varon et al., 2006). Podosomes serve as areas of localized secretion of proteolytic enzymes (Linder, 2007) and couple extracellular adhesion with matrix degradation and promote cellular mobility. For example, macrophages with impaired podosome formation lost migration capacity in a 3D matrix system (Cougoule et al., 2010) and mouse dendritic cells with impaired podosomes lost the ability to migrate across an endothelial monolayer (Calle et al., 2006b).

The ability of podosomes to facilitate ECM degradation is further illustrated in osteoclasts, where podosomes form at the site of bone-matrix contact, and the localized secretion of proteolytic enzymes follows and chews up the bone (Georgess et al., 2014).

The role of podosomes in cell migration was shown to be important in developmental contexts as well. The anchor cells in *C. elegans* embryos have been shown to make actin-rich protrusive structures, which they utilize to breach the basement membrane during development (Hagedorn et al., 2013; Lohmer et al., 2016), and they have been implicated in migration of zebrafish neural crest cells (Murphy et al., 2011). Podosomes are also important in the growth cones for protease localization and ECM degradation during axon cell migration (Santiago-Medina et al., 2015).

In addition to ECM degradation and cell migration, recent evidence is emerging showing that podosomes have mechanosensing ability and can sense and respond to both internal and external cellular forces. Inhibition of myosin II, present at podosome rings, was shown to decrease podosome-exerted traction forces on the underlying substrate (Collin et al., 2008). These traction forces increased with matrix stiffness, demonstrating both internal and external sensing of forces. Flexibility of the substrate can also influence both the turnover and organization of podosomes (Collin et al., 2006). Using micro-patterned surfaces, podosomes in dendritic cells have also been shown to respond to topographic cues (van den Dries et al., 2012). Furthermore, in dendritic cells, podosomes localize to areas of low physical resistance and subsequently facilitate antigen uptake (Baranov et al., 2014). Thus, the function of podosomes is not limited to matrix degradation.

Generally, podosome structure consists of an actin core surrounded by ring of associated proteins. It is generally believed that the polymerization of actin generates the force necessary to push the membrane forward. The actin core contains classic actin polymerization factors, wherein the branched actin polymerizer Arp2/3 is activated by WASP (Linder et al., 2000). The activation of this network is regulated upstream by small RhoGTPases such as Cdc42 (Burns et al., 2001). Recent work has demonstrated the requirement for the small monomeric GTPase ARF1 in podosome induction in multiple cell types by inhibiting RhoA-myosin II actin contractile activity (Rafiq et al., 2017). In addition to branched actin, podosomes in macrophages contain bundled actin along with the actin bundling formin FRL1 (Mersich et al., 2010). Importantly, podosomes in Src-transformed fibroblasts contain the scaffolding protein Tks5 (Abram et al., 2003) presumably providing a mechanism with which to link together various cytoskeletal regulators and polarized membrane transport.

Surrounding the actin ring is a core of associated proteins, such as focal adhesion-associated proteins paxillin, vinculin, and talin (Pfaff and Jurdic, 2001; Gavazzi et al., 1989), which serve to connect the cytoskeleton with integrins at the cell surface. Also present is the mechanosensing protein p130CAS, which is phosphorylated by Src during periods of induced stretching (Alexander et al., 2008; Janostiak et al., 2014). The ring complex connects to the membrane surface via integrins such as β2 and β3, depending on cell type and substrate (Calle et al., 2006a). Additionally, the antigen CD44 has been found in podosomes in both osteoclasts and macrophages where it works to form the actin core and regulate MMP14 localization and subsequent migration (Chabadel et al., 2007; Chellaiah and Ma, 2013).

Podosomes can organize into different shapes, such as tightly contained rosettes, clusters or rings that localize variably throughout the cell, depending on cell type and function. However, all podosomes are connected via acto-myosin filaments that extend out from the core, allowing for communication and the distribution of multiple signals through the entire podosome network (Luxenburg et al., 2007; Meddens et al., 2016). Importantly, podosomes are transient and very dynamic structures with continual turnover of core actin (Destaing et al., 2003). It has been suggested that podosome disassembly is regulated by myosin IIA (Burgstaller and Gimona, 2004). Given that myosin II activity can promote podosome formation in certain contexts (Collin et al., 2006), such a finding highlights the importance of actin dynamics in these structures and the context dependence on their activity. Recently it has been suggested that actin bundling regulates disassembly of podosome, as macro-phages with inhibited fascin demonstrated longer-lived and more stable podosomes (Van Audenhove et al., 2015). This effect was likely due to the opposing role of actin bundling with branched actin nucleation that drives podosome formation.

Improper formation and function of podosomes has been linked to several diseases. The most classic example is Wiskott−Aldrich syndrome, wherein mutations within the WASP gene lead to the loss of podosomes in macro-phages and dendrites, impairing their migratory ability (Linder et al., 1999). The actin cross-linker filamin A has been demonstrated to be required for podosome formation, and podosome misregulation has been suggested to underlie migratory defects during embryonic development that result from filamin A deficiency (Guiet et al., 2012).

Invadopodia and Cancer Cell Metastasis

Similar to podosomes, invadopodia are also actin-rich protrusions of the plasma membrane that extend down into the ECM. They are typically defined by the presence of actin and the scaffolding protein Tks5. Invadopodia are generally the same size and length but have been reported to be more stable, lasting upward of several hours (Gimona et al., 2008). Just like podosomes, invadopodia contain various actin polymerizers and several other proteins such as scaffolding factors, kinases, and various small GTPases that serve to co-ordinate actin polymerization. In contrast to podosomes, which are naturally occurring structures present in multiple cells types, the term invadopodia is limited to such structures that occur in cancer cells and have been reported to occur in several cell types, including breast cancer (Bowden et al., 2006), glioblastoma (Cheerathodi et al., 2016), prostate (Burger et al., 2014), melanoma (Revach et al., 2016), and head and neck squamous cell carcinomas (Jimenez et al., 2015). However, it is clear that podosomes and invadopodia are closely related structures that are likely dependent on a similar machinery governing the actin cytoskeleton and polarized membrane transport during their formation and function.

The primary function of invadopodia is the remodeling of the ECM during cancer cell metastasis. ECM degradation is facilitated by the targeted delivery and secretion of various proteases at the invadopodia. It is through that targeting these proteases to the tip of the extending invadopodia that allows localized ECM degradation and remodeling, thus leading to invasion of cancer cells. The most prominent and well-studied group are the matrix metalloproteinases (MMPs), in particular MT1-MMP (MMP14), MMP2, and MMP9 (Jacob and Prekeris, 2015). MMPs are the large family of proteins that degrade a wide variety of ECM components, such as collagen, laminin, gelatin, and fibronectin. Also important are the A Disintegrin And Metalloproteinase (ADAM) family of proteases, especially ADAM12. ADAM12 acts to shed membrane bound proteins from the cell surface proteins and activate growth factors (Shi et al., 2000; Asakura et al., 2002) and thereby induce invadopodia formation (Diaz et al., 2013). Additionally, the serine proteases FAPα and DPP4 have been suggested to localize to invadopodia and regulate matrix degradation in HUVEC cells (Ghersi et al., 2006).

Similar to podosomes, recent work has highlighted the mechanosensing capabilities of invadopodia. Work has shown that increased stiffness of the ECM leads to an increase in invadopodia and their degradative capacity (Alexander et al., 2008; Parekh et al., 2011; Aung et al., 2014) and is thought to facilitate these effects via myosin II and p130CAS (Alexander et al., 2008). Additionally, interaction and adhesion of invadopodia to the ECM has been shown to be important and mediated largely through the various β1 integrin heterodimers (Coopman et al., 1996) as well as αVβ3 integrin (Malik et al., 2010).

Unlike podosomes, which occur naturally, invadopodia are instead induced in cancer cells. The most common method of invadopodia induction is stimulation by external growth factors. Perhaps the most common method of induction is through activation of receptor tyrosine kinases (RTKs). Stimulation by EGF (Mader et al., 2011), HGF (Rajadurai et al., 2012) and TGFβ (Varon et al., 2006) have all been shown to lead to the formation of matrix-degrading invadopodia in multiple cell types. In addition to RTK signaling, cellular stresses such as hypoxia (Lucien et al., 2011) and reactive oxygen species (Diaz et al., 2009) have also been shown to lead to an induction of invadopodia formation.

In general, extracellular stimulation leads to activation of the tyrosine kinase Src. Activated Src then phosphorylates the scaffolding protein Tks5, leading to Tks5 binding to the plasma membrane (Abram et al., 2003; Courtneidge et al., 2005) at sites enriched in phosphatidylinositol PI(3,4)P2. Tks5 can then recruit (Crimaldi et al., 2009) and Src can phosphorylate (Tehrani et al., 2007) the actin polymerizing protein cortactin, leading to complex formation by recruiting other actin regulators such as N-WASP and ARP2/3 (Stengel and Zheng, 2011) and the formation of branched actin networks that generate the necessary force for invadopodia to protrude into the basement membrane.

Additionally, several invadopodia inducing signals also lead to activation of Rho family GTPase Cdc42 (Stengel and Zheng, 2011), which can interact with WASP to initiate rearrangement of the actin cytoskeleton (Rohatgi et al., 2000). Constitutively active Cdc42 can induce invadopodia formation (Razidlo et al., 2014) and, conversely, downregulation of Cdc42 has been shown to inhibit invasion and ECM degradation (Desmarais et al., 2009). Relative to invadopodia, several GEFs have been shown to activate Cdc42, such as Vav1 (Razidlo et al., 2014) and Fdg1 (Ayala et al., 2009). However, the full complement of upstream GAPs and GEFs in any given system that lead to Cdc42 activation remains to be fully defined.

It has become clear that invadopodia are highly dynamic and usually undergoes several cycles of extension and retraction during cell migration. However, little is known about how invadopodia turnover and disassemble. Recently it was shown that the Rho GTPase Rac1 activation by Trio-GEF leads to turnover of invadopodia in metastatic breast carcinoma cells (Moshfegh et al., 2014). This occurs independent of and in contrast to Rac1 promigratory and proinvasive functions, highlighting the spatiotemporal dynamics that govern these structures. Recent work has shown that the ubiquitin ligase scaffold Cul5 can mediate ubiquitination of p-Src (Laszlo and Cooper, 2009), resulting in p-Src degradation, a decrease in cellular transformation and altered membrane dynamics (Teckchandani et al., 2014). Cul5 has also been shown to downregulate various small GTPases known to regulate invadopodia protrusion, such as Vav1 (Okumura et al., 2016) and may be a novel regulator of invadopodia turnover. Cullins typically bind to and recruit-specific ubiquitin ligases to ensure specificity of target protein ubiquitination. However, that ubiquitin ligase binds Cul5 in the invadopodia remains to be determined. It is also unclear how Cul5 itself is targeted to invadopodia and what regulates its activation and function during retraction of invadopodia. Consequently, further studies will need to explore the mechanisms of invadopodia retraction and dynamics.

Membrane Transport and MMP Targeting During Invadopodia Formation

A fundamental step in cancer progression is the breakdown of the ECM, a process that depends on the targeting of matrix-degrading proteolytic enzymes to basal surfaces of cells (Kessenbrock et al., 2010). This remodeling of the ECM both releases growth factors utilized by the invading cells and provides space through which these cells will migrate (Wolf et al., 2007). The major class of proteolytic enzymes is MMPs, a family of calcium-dependent zinc endopeptidases that degrade ECM proteins. They are broadly categorized based on substrate specificity, such as the gelatinases MMP2 and MMP9 and collagenases such as MMP14 (Khokha et al., 2013). These three MMPs have been shown to be crucial for localized ECM degradation in cancer metastasis and are targeted to the invadopodia in cancer cells. While the mechanisms that govern the targeting and secretion of MMPs at the invadopodia remains to be understood here we will summarize the newest findings about the regulation of MMP14 and MMP2/9 targeting during cancer cell invasion.

MMP14 Targeting During Invadopodia Formation

MMP14, also called MT1-MMP, is a type I transmembrane protein capable of cleaving gelatin, fibronectin, and laminin and is also required for processing and activation of MMP2 (Sato et al., 1994). In addition, it is required for basement membrane degradation and is thus central to the invasive process. Localization of MMP14 to the invasive front is typically triggered by β1 integrin adhesion (Wolf et al., 2003; Ellerbroek et al., 2001). Initially inactive, it is processed on transport to the plasma membrane. On delivery to invadopodia plasma membrane, it is quickly internalized and is either shuttled to the lysosome for degradation or recycled back to the surface. Internalization of MMP14 is regulated by both clathrin-mediated endocytosis (Uekita et al., 2001) and by caveolae-mediated endocytosis (Annabi et al., 2004). It is generally accepted that this internalization is what limits the activity of MMP14 by sequestering it within intracellular endocytic compartments (Fig. 5.3A).

After internalization, several mechanisms have been demonstrated to traffic MMP14 back to the cell surface. The small monomeric GTPase Rab8, initially shown to regulate vesicle trafficking between the TGN and the basolateral surface (Huber et al., 1993), has been shown to regulate MMP14 exocytosis from internal endocytic storage compartments within the invadopodia in invasive breast cancer cells (Bravo-Cordero et al., 2007). Additionally, Rab5a and Rab4 GTPases promote exocytosis of MMP14 presumably by regulating its transport via sorting and recycling endosomes (Frittoli et al., 2014). Finally, Rab2a, previously demonstrated to regulate ER-Golgi trafficking (Ni et al., 2002), has been shown to interact with the late-endosomal protein VSP39 to mediate MMP14 secretion and matrix degradation (Kajiho et al., 2016). Similarly, the late-endosomal marker Rab7 regulates MMP14 delivery, in conjunction with vesicle-associated membrane protein 7 in invasive breast cancer cells (Williams and Coppolino, 2011; Steffen et al., 2008). What remains unclear is how all these well-established regulators of endocytic

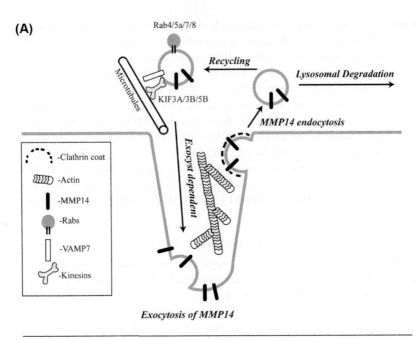

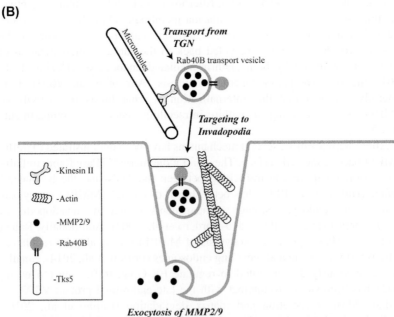

FIGURE 5.3 (A) MMP14 at the plasma membrane is endocytosed and either sent to the lysosome for degradation or recycled back to the membrane via Exocyst-dependent vesicle trafficking. Several Rab GTPases have been shown to regulate this process. Trafficking along microtubules has been shown to be facilitated by multiple kinesin motors. (B) Rab40b vesicles carrying MMP2 and MMP9 are trafficked along microtubules from the TGN to invadopodia. Rab40b binds to invadopodia precursor Tks5 and subsequently releases MMP cargo.

membrane transport cross talk to sequentially regulate MMP14 internalization and recycling back to the invadopodia plasma membrane (Fig. 5.3A).

While the molecular machinery governing the internalization and trafficking of MMP14 through endocytic organelles is beginning to be established, what targets MMP14 transport vesicles to the invadopodia is less understood. Typically, transport vesicles rely on specific tethering factors to recruit them to the site of exocytosis. Significantly, the Exocyst complex, a well-established plasma membrane tethering factor has recently been shown to be important in regulating MMP14 delivery and invadopodia formation (Liu et al., 2009; Monteiro et al., 2013). Thus, it was proposed that the Exocyst complex ensures the targeted delivery of MMP14 to the invadopodia by binding and tethering MMP14-vesicles (Fig. 5.3A). However, since the Exocyst complex is also present outside the invadopodia, the additional factors are likely needed to ensure the fidelity of MMP14 targeting. It has been shown that the Exocyst complex function is dependent on the Arp2/3-activator Wiskott—Aldrich syndrome protein and Scar homolog WASH, which itself is involved in endosomal trafficking (Gomez et al., 2012). Since WASH and Arp2/3 are all enriched in the invadopodia, it is possible that the Exocyst complex and Arp2/3-dependent branched actin cytoskeleton work together to target MMP14 during invadopodia formation and extension.

In addition to tethering factors, the molecular motors that mediate vesicle trafficking along actin filaments and microtubules also participate in regulating protein targeting. Consequently, kinesins that regulate the trafficking of MMP14 along microtubules are beginning to be identified. The plus-end directed motors KIF5B and KIF3A/KIF3B are both necessary for MMP14 transport in macrophages (Wiesner et al., 2010). Additionally, the KIF5 adapter proteins JIP3 and JIP4 can regulate MMP14 delivery to invadopodia in breast cancer cells (Marchesin et al., 2015). Thus, it is likely that coordinated function of kinesins, actin regulators and the Exocyst tethering complex all contribute to ensure the delivery of MMP14 to the invadopodia during cancer metastasis (Fig. 5.3A).

MMP2 and MMP9 Targeting During Invadopodia Formation

In addition to MMP14, two other MMPs, MMP2, and MMP9 have been shown to be influential in matrix degradation during cancer cell invasion and have increased localization at invadopodia. Like MMP14, MMP2, and MMP9 are initially secreted in an inactive form and are only catalytically active after processing by other MMPs (Sato et al., 1994; Ramos-DeSimone et al., 1999) Although the trafficking of MMP2 and MMP9 is highly studied, much less is known about the mechanisms that govern their trafficking to invadopodia. However, some details are starting to emerge. Similar to MMP14, the actin cytoskeleton and cortactin are also required for the delivery and secretion of MMP2 and MMP9 (Clark et al., 2007). Loss of the cytosolic protein Secernin 1, which is a regulator of exocytosis in mast cells (Way et al., 2002), was found to decrease secretion of both MMP2 and MMP9 in colon cancer cells (Lin et al., 2015) and inhibited subsequent cellular invasion. However, the mechanism behind this regulation is unclear. In mice, knockout of the exocytosis regulator Rab11 (Takahashi et al., 2012) significantly reduced the secretion of both MMP2 and MMP9, although this has yet to be shown in the context of invadopodia (Yu et al., 2014). Like MMP14, MMP2 and MMP9-containing vesicles are also trafficked along microtubule networks by the motor protein kinesin (Schnaeker et al., 2004) (Fig. 5.3B). Significantly, it was shown that MMP2 and MMP9 secretion neither depend on canonical endocytic trafficking regulators, such as Rab5 or Rab4, nor does it rely on the Exocyst complex (Jacob et al., 2013), implying that MMP2/9 and MMP14 are transported and targeted via different molecular mechanisms. Consistent with this hypothesis, recent work has shown that in breast cancer cells Rab40b GTPase regulates the trafficking of MMP2 and MMP9 to invadopodia and is required for breast cancer invasion in vitro and metastasis in vivo (Jacob et al., 2016) (Fig. 5.3B). Additionally, Rab40b has been shown to bind to the invadopodia precursor protein Tks5 (Jacob et al., 2016). What makes this discovery especially exciting is the fact that Tks5 is a known metastasis regulator that is highly enriched in invadopodia (Crimaldi et al., 2009). Thus, a Tks5 and Rab40b complex is perfectly suited to play the role of an invadopodia-specific tether that targets MMP2/9 transport vesicles to the invadopodia during cancer cell invasion (Fig. 5.3B). What remains to be determined is how the formation of Rab40b and Tks5 complex is regulated in the context of the invadopodia initiation, extension, and retraction. Further studies will be required to define these regulatory pathways and to determine whether these pathways may also play a role in podosome formation as well as cell migration during development.

CONCLUDING REMARKS

It has becoming evident that coordinated interplay between post-Golgi membrane transport and cytoskeleton dynamics play a major role during the establishment of epithelial polarity. Significantly, the basic machinery regulating this polarized membrane transport is similar in cells that terminally polarize as well as cell that form transient polarized structures. However, many questions remain. We still understand little about the machinery governing coordinated regulation of endocytic membranes and cytoskeleton. Even less is known about how these changes in membrane and cytoskeleton are regulated in vivo since most of the studies identifying polarization machinery have been done in vitro. Thus, future studies will be needed to further define the roles of actin, microtubules, and membrane transport during cell polarization in different tissues and developmental contexts.

ACKNOWLEDGMENTS

We are grateful to Peter Dempsey (University of Colorado AMC) for critical reading of the manuscript. We apologize to all colleagues whose work could not be cited due to space limitations. The work in Dr. Rytis Prekeris laboratory is supported by support by NIH grant DK064380 (to R.P.), grant from Cancer League of Colorado (to R.P.), and NSF Graduate Research Fellowship (to CEJ, grant number DGE-1553798).

REFERENCES

Abram, C.L., Seals, D.F., et al., 2003. The adaptor protein fish associates with members of the ADAMs family and localizes to podosomes of Src-transformed cells. J. Biol. Chem. 278 (19), 16844–16851.

Alexander, N.R., Branch, K.M., et al., 2008. Extracellular matrix rigidity promotes invadopodia activity. Curr. Biol. 18 (17), 1295–1299.

Alvers, A.L., Ryan, S., et al., 2014. Single continuous lumen formation in the zebrafish gut is mediated by smoothened-dependent tissue remodeling. Development 141 (5), 1110–1119.

Annabi, B., Thibeault, S., et al., 2004. Hyaluronan cell surface binding is induced by type I collagen and regulated by caveolae in glioma cells. J. Biol. Chem. 279 (21), 21888–21896.

Apodaca, G., Gallo, L.I., et al., 2012. Role of membrane traffic in the generation of epithelial cell asymmetry. Nat. Cell Biol. 14 (12), 1235–1243.

Armenti, S.T., Chan, E., et al., 2014. Polarized exocyst-mediated vesicle fusion directs intracellular lumenogenesis within the *C. elegans* excretory cell. Dev. Biol. 394 (1), 110–121.

Asakura, M., Kitakaze, M., et al., 2002. Cardiac hypertrophy is inhibited by antagonism of ADAM12 processing of HB-EGF: metalloproteinase inhibitors as a new therapy. Nat. Med. 8 (1), 35–40.

Assemat, E., Bazellieres, E., et al., 2008. Polarity complex proteins. Biochim. Biophys. Acta 1778 (3), 614–630.

Aung, A., Seo, Y.N., et al., 2014. 3D traction stresses activate protease-dependent invasion of cancer cells. Biophys. J. 107 (11), 2528–2537.

Ayala, I., Giacchetti, G., et al., 2009. Faciogenital dysplasia protein Fgd1 regulates invadopodia biogenesis and extracellular matrix degradation and is up-regulated in prostate and breast cancer. Cancer Res. 69 (3), 747–752.

Baetz, N.W., Goldenring, J.R., 2013. Rab11-family interacting proteins define spatially and temporally distinct regions within the dynamic Rab11a-dependent recycling system. Mol. Biol. Cell 24 (5), 643–658.

Baranov, M.V., Ter Beest, M., et al., 2014. Podosomes of dendritic cells facilitate antigen sampling. J. Cell Sci. 127 (Pt 5), 1052–1064.

Bossinger, O., Klebes, A., et al., 2001. Zonula adherens formation in *Caenorhabditis elegans* requires dlg-1, the homologue of the *Drosophila* gene discs large. Dev. Biol. 230 (1), 29–42.

Bowden, E.T., Onikoyi, E., et al., 2006. Co-localization of cortactin and phosphotyrosine identifies active invadopodia in human breast cancer cells. Exp. Cell Res. 312 (8), 1240–1253.

Bravo-Cordero, J.J., Marrero-Diaz, R., et al., 2007. MT1-MMP proinvasive activity is regulated by a novel Rab8-dependent exocytic pathway. EMBO J. 26 (6), 1499–1510.

Burger, K.L., Learman, B.S., et al., 2014. Src-dependent Tks5 phosphorylation regulates invadopodia-associated invasion in prostate cancer cells. Prostate 74 (2), 134–148.

Burgstaller, G., Gimona, M., 2004. Actin cytoskeleton remodelling via local inhibition of contractility at discrete microdomains. J. Cell Sci. 117 (Pt 2), 223—231.

Burns, S., Thrasher, A.J., et al., 2001. Configuration of human dendritic cell cytoskeleton by Rho GTPases, the WAS protein, and differentiation. Blood 98 (4), 1142—1149.

Calle, Y., Burns, S., et al., 2006a. The leukocyte podosome. Eur. J. Cell Biol. 85 (3—4), 151—157.

Calle, Y., Carragher, N.O., et al., 2006b. Inhibition of calpain stabilises podosomes and impairs dendritic cell motility. J. Cell Sci. 119 (Pt 11), 2375—2385.

Casanova, J.E., Wang, X., et al., 1999. Association of Rab25 and Rab11a with the apical recycling system of polarized Madin-Darby canine kidney cells. Mol. Biol. Cell 10 (1), 47—61.

Chabadel, A., Banon-Rodriguez, I., et al., 2007. CD44 and beta3 integrin organize two functionally distinct actin-based domains in osteoclasts. Mol. Biol. Cell 18 (12), 4899—4910.

Chavrier, P., Vingron, M., et al., 1990. Molecular cloning of YPT1/SEC4-related cDNAs from an epithelial cell line. Mol. Cell Biol. 10 (12), 6578—6585.

Cheerathodi, M., Avci, N.G., et al., 2016. The cytoskeletal adapter protein spinophilin regulates invadopodia dynamics and tumor cell invasion in glioblastoma. Mol. Cancer Res. 14 (12), 1277—1287.

Chellaiah, M.A., Ma, T., 2013. Membrane localization of membrane type 1 matrix metalloproteinase by CD44 regulates the activation of pro-matrix metalloproteinase 9 in osteoclasts. Biomed. Res. Int. 2013, 302392.

Chen, W.T., Chen, J.M., et al., 1985. Local degradation of fibronectin at sites of expression of the transforming gene product pp60src. Nature 316 (6024), 156—158.

Chen, W.T., 1989. Proteolytic activity of specialized surface protrusions formed at rosette contact sites of transformed cells. J. Exp. Zool. 251 (2), 167—185.

Clark, E.S., Whigham, A.S., et al., 2007. Cortactin is an essential regulator of matrix metalloproteinase secretion and extracellular matrix degradation in invadopodia. Cancer Res. 67 (9), 4227—4235.

Collin, O., Tracqui, P., et al., 2006. Spatiotemporal dynamics of actin-rich adhesion microdomains: influence of substrate flexibility. J. Cell Sci. 119 (Pt 9), 1914—1925.

Collin, O., Na, S., et al., 2008. Self-organized podosomes are dynamic mechanosensors. Curr. Biol. 18 (17), 1288—1294.

Coopman, P.J., Thomas, D.M., et al., 1996. Integrin alpha 3 beta 1 participates in the phagocytosis of extracellular matrix molecules by human breast cancer cells. Mol. Biol. Cell 7 (11), 1789—1804.

Cougoule, C., Le Cabec, V., et al., 2010. Three-dimensional migration of macrophages requires Hck for podosome organization and extracellular matrix proteolysis. Blood 115 (7), 1444—1452.

Courtneidge, S.A., Azucena, E.F., et al., 2005. The SRC substrate Tks5, podosomes (invadopodia), and cancer cell invasion. Cold Spring Harb. Symp. Quant. Biol. 70, 167—171.

Crimaldi, L., Courtneidge, S.A., et al., 2009. Tks5 recruits AFAP-110, p190RhoGAP, and cortactin for podosome formation. Exp. Cell Res. 315 (15), 2581—2592.

Das, S., Hehnly, H., et al., 2014. A new role for Rab GTPases during early mitotic stages. Small GTPases 5, e29565.

Datta, A., Bryant, D.M., et al., 2011. Molecular regulation of lumen morphogenesis. Curr. Biol. 21 (3), R126—R136.

Desmarais, V., Yamaguchi, H., et al., 2009. N-WASP and cortactin are involved in invadopodium-dependent chemotaxis to EGF in breast tumor cells. Cell Motil. Cytoskelet. 66 (6), 303—316.

Destaing, O., Saltel, F., et al., 2003. Podosomes display actin turnover and dynamic self-organization in osteoclasts expressing actin-green fluorescent protein. Mol. Biol. Cell 14 (2), 407—416.

Diaz, B., Shani, G., et al., 2009. Tks5-dependent, nox-mediated generation of reactive oxygen species is necessary for invadopodia formation. Sci. Signal 2 (88), ra53.

Diaz, B., Yuen, A., et al., 2013. Notch increases the shedding of HB-EGF by ADAM12 to potentiate invadopodia formation in hypoxia. J. Cell Biol. 201 (2), 279—292.

Ellerbroek, S.M., Wu, Y.I., et al., 2001. Functional interplay between type I collagen and cell surface matrix metalloproteinase activity. J. Biol. Chem. 276 (27), 24833—24842.

Erickson, R.P., Larson-Thome, K., et al., 2008. Navajo microvillous inclusion disease is due to a mutation in MYO5B. Am. J. Med. Genet. A 146A (24), 3117—3119.

Frittoli, E., Palamidessi, A., et al., 2014. A RAB5/RAB4 recycling circuitry induces a proteolytic invasive program and promotes tumor dissemination. J. Cell Biol. 206 (2), 307—328.

Galvez-Santisteban, M., Rodriguez-Fraticelli, A.E., et al., 2012. Synaptotagmin-like proteins control the formation of a single apical membrane domain in epithelial cells. Nat. Cell Biol. 14 (8), 838—849.

Gavazzi, I., Nermut, M.V., et al., 1989. Ultrastructure and gold-immunolabelling of cell-substratum adhesions (podosomes) in RSV-transformed BHK cells. J. Cell Sci. 94 (Pt 1), 85—99.

Georgess, D., Machuca-Gayet, I., et al., 2014. Podosome organization drives osteoclast-mediated bone resorption. Cell Adhes. Migr. 8 (3), 191—204.

Ghersi, G., Zhao, Q., et al., 2006. The protease complex consisting of dipeptidyl peptidase IV and seprase plays a role in the migration and invasion of human endothelial cells in collagenous matrices. Cancer Res. 66 (9), 4652—4661.

Gimona, M., Buccione, R., et al., 2008. Assembly and biological role of podosomes and invadopodia. Curr. Opin. Cell Biol. 20 (2), 235—241.

Goldenring, J.R., Shen, K.R., et al., 1993. Identification of a small GTP-binding protein, Rab25, expressed in the gastrointestinal mucosa, kidney, and lung. J. Biol. Chem. 268 (25), 18419—18422.

Gomez, T.S., Gorman, J.A., et al., 2012. Trafficking defects in WASH-knockout fibroblasts originate from collapsed endosomal and lysosomal networks. Mol. Biol. Cell 23 (16), 3215—3228.

Guiet, R., Verollet, C., et al., 2012. Macrophage mesenchymal migration requires podosome stabilization by filamin A. J. Biol. Chem. 287 (16), 13051—13062.

Hagedorn, E.J., Ziel, J.W., et al., 2013. The netrin receptor DCC focuses invadopodia-driven basement membrane transmigration in vivo. J. Cell Biol. 201 (6), 903—913.

Hales, C.M., Griner, R., et al., 2001. Identification and characterization of a family of Rab11-interacting proteins. J. Biol. Chem. 276 (42), 39067—39075.

Hales, C.M., Vaerman, J.P., et al., 2002. Rab11 family interacting protein 2 associates with Myosin Vb and regulates plasma membrane recycling. J. Biol. Chem. 277 (52), 50415—50421.

Hehnly, H., Doxsey, S., 2014. Rab11 endosomes contribute to mitotic spindle organization and orientation. Dev. Cell 28 (5), 497—507.

Horne-Badovinac, S., Lin, D., et al., 2001. Positional cloning of heart and soul reveals multiple roles for PKC lambda in zebrafish organogenesis. Curr. Biol. 11 (19), 1492—1502.

Hsu, S.C., TerBush, D., et al., 2004. The exocyst complex in polarized exocytosis. Int. Rev. Cytol. 233, 243—265.

Huber, L.A., Pimplikar, S., et al., 1993. Rab8, a small GTPase involved in vesicular traffic between the TGN and the basolateral plasma membrane. J. Cell Biol. 123 (1), 35−45.

Hutagalung, A.H., Novick, P.J., 2011. Role of Rab GTPases in membrane traffic and cell physiology. Physiol. Rev. 91 (1), 119−149.

Jacob, A., Prekeris, R., 2015. The regulation of MMP targeting to invadopodia during cancer metastasis. Front. Cell Dev. Biol. 3, 4.

Jacob, A., Jing, J., et al., 2013. Rab40b regulates trafficking of MMP2 and MMP9 during invadopodia formation and invasion of breast cancer cells. J. Cell Sci. 126 (Pt 20), 4647−4658.

Jacob, A., Linklater, E., et al., 2016. The role and regulation of Rab40b-Tks5 complex during invadopodia formation and cancer cell invasion. J. Cell Sci. 129 (23), 4341−4353.

Janostiak, R., Pataki, A.C., et al., 2014. Mechanosensors in integrin signaling: the emerging role of p130Cas. Eur. J. Cell Biol. 93 (10−12), 445−454.

Jimenez, L., Sharma, V.P., et al., 2015. MicroRNA-375 suppresses extracellular matrix degradation and invadopodial activity in head and neck squamous cell carcinoma. Arch. Pathol. Lab. Med. 139 (11), 1349−1361.

Jing, J., Prekeris, R., 2009. Polarized endocytic transport: the roles of Rab11 and Rab11-FIPs in regulating cell polarity. Histol. Histopathol. 24 (9), 1171−1180.

Jones, T.A., Metzstein, M.M., 2011. A novel function for the PAR complex in subcellular morphogenesis of tracheal terminal cells in *Drosophila melanogaster*.

Jones, T.A., Nikolova, L.S., et al., 2014. Exocyst-mediated membrane trafficking is required for branch outgrowth in *Drosophila* tracheal terminal cells. Dev. Biol. 390 (1), 41−50.

Junutula, J.R., Schonteich, E., et al., 2004. Molecular characterization of Rab11 interactions with members of the family of Rab11-interacting proteins. J. Biol. Chem. 279 (32), 33430−33437.

Kajiho, H., Kajiho, Y., et al., 2016. RAB2A controls MT1-MMP endocytic and E-cadherin polarized Golgi trafficking to promote invasive breast cancer programs. EMBO Rep. 17 (7), 1061−1080.

Kaverina, I., Stradal, T.E., et al., 2003. Podosome formation in cultured A7r5 vascular smooth muscle cells requires Arp2/3-dependent de-novo actin polymerization at discrete microdomains. J. Cell Sci. 116 (Pt 24), 4915−4924.

Kessenbrock, K., Plaks, V., et al., 2010. Matrix metalloproteinases: regulators of the tumor microenvironment. Cell 141 (1), 52−67.

Khokha, R., Murthy, A., et al., 2013. Metalloproteinases and their natural inhibitors in inflammation and immunity. Nat. Rev. Immunol. 13 (9), 649−665.

Klinkert, K., Rocancourt, M., et al., 2016. Rab35 GTPase couples cell division with initiation of epithelial apico-basal polarity and lumen opening. Nat. Commun. 7, 11166.

Knodler, A., Feng, S., et al., 2010. Coordination of Rab8 and Rab11 in primary ciliogenesis. Proc. Natl. Acad. Sci. U.S.A. 107 (14), 6346−6351.

Knowles, B.C., Roland, J.T., et al., 2014. Myosin Vb uncoupling from RAB8A and RAB11A elicits microvillus inclusion disease. J. Clin. Investig. 124 (7), 2947−2962.

Lapierre, L.A., Kumar, R., et al., 2001. Myosin vb is associated with plasma membrane recycling systems. Mol. Biol. Cell 12 (6), 1843−1857.

Lapierre, L.A., Dorn, M.C., et al., 2003. Rab11b resides in a vesicular compartment distinct from Rab11a in parietal cells and other epithelial cells. Exp. Cell Res. 290 (2), 322−331.

Laszlo, G.S., Cooper, J.A., 2009. Restriction of src activity by Cullin-5. Curr. Biol. 19 (2), 157−162.

Li, B.X., Satoh, A.K., et al., 2007. Myosin V, Rab11, and dRip11 direct apical secretion and cellular morphogenesis in developing *Drosophila* photoreceptors. J. Cell Biol. 177 (4), 659−669.

Li, D., Mangan, A., et al., 2014. FIP5 phosphorylation during mitosis regulates apical trafficking and lumenogenesis. EMBO Rep. 15 (4), 428—437.

Lin, S., Jiang, T., et al., 2015. Secernin-1 contributes to colon cancer progression through enhancing matrix metalloproteinase-2/9 exocytosis. Dis. Markers 2015, 230703.

Linder, S., Nelson, D., et al., 1999. Wiskott-Aldrich syndrome protein regulates podosomes in primary human macrophages. Proc. Natl. Acad. Sci. U.S.A. 96 (17), 9648—9653.

Linder, S., Higgs, H., et al., 2000. The polarization defect of Wiskott-Aldrich syndrome macrophages is linked to dislocalization of the Arp2/3 complex. J. Immunol. 165 (1), 221—225.

Linder, S., 2007. The matrix corroded: podosomes and invadopodia in extracellular matrix degradation. Trends Cell Biol. 17 (3), 107—117.

Liu, J., Yue, P., et al., 2009. The role of the exocyst in matrix metalloproteinase secretion and actin dynamics during tumor cell invadopodia formation. Mol. Biol. Cell 20 (16), 3763—3771.

Lohmer, L.L., Clay, M.R., et al., 2016. A sensitized screen for genes promoting invadopodia function in vivo: CDC-42 and Rab GDI-1 direct distinct aspects of invadopodia formation. PLoS Genet. 12 (1), e1005786.

Lu, R., Wilson, J.M., 2016. Rab14 specifies the apical membrane through Arf6-mediated regulation of lipid domains and Cdc42. Sci. Rep. 6, 38249.

Lucien, F., Brochu-Gaudreau, K., et al., 2011. Hypoxia-induced invadopodia formation involves activation of NHE-1 by the p90 ribosomal S6 kinase (p90RSK). PLoS One 6 (12), e28851.

Luxenburg, C., Geblinger, D., et al., 2007. The architecture of the adhesive apparatus of cultured osteoclasts: from podosome formation to sealing zone assembly. PLoS One 2 (1), e179.

Mader, C.C., Oser, M., et al., 2011. An EGFR-Src-Arg-cortactin pathway mediates functional maturation of invadopodia and breast cancer cell invasion. Cancer Res. 71 (5), 1730—1741.

Malik, G., Knowles, L.M., et al., 2010. Plasma fibronectin promotes lung metastasis by contributions to fibrin clots and tumor cell invasion. Cancer Res. 70 (11), 4327—4334.

Mancuso, V.P., Parry, J.M., et al., 2012. Extracellular leucine-rich repeat proteins are required to organize the apical extracellular matrix and maintain epithelial junction integrity in *C. elegans*. Development 139 (5), 979—990.

Mangan, A.J., Sietsema, D.V., et al., 2016. Cingulin and actin mediate midbody-dependent apical lumen formation during polarization of epithelial cells. Nat. Commun. 7, 12426.

Marchesin, V., Castro-Castro, A., et al., 2015. ARF6-JIP3/4 regulate endosomal tubules for MT1-MMP exocytosis in cancer invasion. J. Cell Biol. 211 (2), 339—358.

Mazelova, J., Astuto-Gribble, L., et al., 2009. Ciliary targeting motif VxPx directs assembly of a trafficking module through Arf4. EMBO J. 28 (3), 183—192.

Meddens, M.B., Pandzic, E., et al., 2016. Actomyosin-dependent dynamic spatial patterns of cytoskeletal components drive mesoscale podosome organization. Nat. Commun. 7, 13127.

Mersich, A.T., Miller, M.R., et al., 2010. The formin FRL1 (FMNL1) is an essential component of macrophage podosomes. Cytoskelet. (Hoboken) 67 (9), 573—585.

Monteiro, P., Rosse, C., et al., 2013. Endosomal WASH and exocyst complexes control exocytosis of MT1-MMP at invadopodia. J. Cell Biol. 203 (6), 1063—1079.

Moshfegh, Y., Bravo-Cordero, J.J., et al., 2014. A Trio-Rac1-Pak1 signalling axis drives invadopodia disassembly. Nat. Cell Biol. 16 (6), 574—586.

Mrozowska, P.S., Fukuda, M., 2016. Regulation of podocalyxin trafficking by Rab small GTPases in 2D and 3D epithelial cell cultures. J. Cell Biol. 213 (3), 355—369.

Muller, T., Hess, M.W., et al., 2008. MYO5B mutations cause microvillus inclusion disease and disrupt epithelial cell polarity. Nat. Genet. 40 (10), 1163—1165.

Murphy, D.A., Diaz, B., et al., 2011. A Src-Tks5 pathway is required for neural crest cell migration during embryonic development. PLoS One 6 (7), e22499.

Ng, A.N., de Jong-Curtain, T.A., et al., 2005. Formation of the digestive system in zebrafish: III. Intestinal epithelium morphogenesis. Dev. Biol. 286 (1), 114−135.

Ni, X., Ma, Y., et al., 2002. Molecular cloning and characterization of a novel human Rab (Rab2B) gene. J. Hum. Genet. 47 (10), 548−551.

Novick, P., 2016. Regulation of membrane traffic by Rab GEF and GAP cascades. Small GTPases 7 (4), 252−256.

Okumura, F., Joo-Okumura, A., et al., 2016. The role of cullin 5-containing ubiquitin ligases. Cell Div. 11, 1.

Parekh, A., Ruppender, N.S., et al., 2011. Sensing and modulation of invadopodia across a wide range of rigidities. Biophys. J. 100 (3), 573−582.

Pfaff, M., Jurdic, P., 2001. Podosomes in osteoclast-like cells: structural analysis and cooperative roles of paxillin, proline-rich tyrosine kinase 2 (Pyk2) and integrin alphaVbeta3. J. Cell Sci. 114 (Pt 15), 2775−2786.

Prekeris, R., 2003. Rabs, Rips, FIPs, and endocytic membrane traffic. ScientificWorldJournal 3, 870−880.

Rafiq, N.B., Lieu, Z.Z., et al., 2017. Podosome assembly is controlled by the GTPase ARF1 and its nucleotide exchange factor ARNO. J. Cell Biol. 216 (1), 181−197.

Rajadurai, C.V., Havrylov, S., et al., 2012. Met receptor tyrosine kinase signals through a cortactin-Gab1 scaffold complex, to mediate invadopodia. J. Cell Sci. 125 (Pt 12), 2940−2953.

Ramos-DeSimone, N., Hahn-Dantona, E., et al., 1999. Activation of matrix metalloproteinase-9 (MMP-9) via a converging plasmin/stromelysin-1 cascade enhances tumor cell invasion. J. Biol. Chem. 274 (19), 13066−13076.

Razidlo, G.L., Schroeder, B., et al., 2014. Vav1 as a central regulator of invadopodia assembly. Curr. Biol. 24 (1), 86−93.

Revach, O.Y., Winograd-Katz, S.E., et al., 2016. The involvement of mutant Rac1 in the formation of invadopodia in cultured melanoma cells. Exp. Cell Res. 343 (1), 82−88.

Rink, J., Ghigo, E., et al., 2005. Rab conversion as a mechanism of progression from early to late endosomes. Cell 122 (5), 735−749.

Rohatgi, R., Ho, H.Y., et al., 2000. Mechanism of N-WASP activation by CDC42 and phosphatidylinositol 4, 5-bisphosphate. J. Cell Biol. 150 (6), 1299−1310.

Roland, J.T., Bryant, D.M., et al., 2011. Rab GTPase-Myo5B complexes control membrane recycling and epithelial polarization. Proc. Natl. Acad. Sci. U.S.A. 108 (7), 2789−2794.

Roman-Fernandez, A., Bryant, D.M., 2016. Complex polarity: building multicellular tissues through apical membrane traffic. Traffic 17 (12), 1244−1261.

Santiago-Medina, M., Gregus, K.A., et al., 2015. Regulation of ECM degradation and axon guidance by growth cone invadosomes. Development 142 (3), 486−496.

Sato, H., Takino, T., et al., 1994. A matrix metalloproteinase expressed on the surface of invasive tumour cells. Nature 370 (6484), 61−65.

Schiel, J.A., Childs, C., et al., 2013. Endocytic transport and cytokinesis: from regulation of the cytoskeleton to midbody inheritance. Trends Cell Biol. 23 (7), 319−327.

Schnaeker, E.M., Ossig, R., et al., 2004. Microtubule-dependent matrix metalloproteinase-2/matrix metalloproteinase-9 exocytosis: prerequisite in human melanoma cell invasion. Cancer Res. 64 (24), 8924−8931.

Schottenfeld-Roames, J., Ghabrial, A.S., 2012. Whacked and Rab35 polarize dynein-motor-complex-dependent seamless tube growth. Nat. Cell Biol. 14 (4), 386−393.

Shi, Z., Xu, W., et al., 2000. ADAM 12, a disintegrin metalloprotease, interacts with insulin-like growth factor-binding protein-3. J. Biol. Chem. 275 (24), 18574−18580.

Sobajima, T., Yoshimura, S., et al., 2014. Rab11a is required for apical protein localisation in the intestine. Biol. Open 4 (1), 86–94.

Steffen, A., Le Dez, G., et al., 2008. MT1-MMP-dependent invasion is regulated by TI-VAMP/VAMP7. Curr. Biol. 18 (12), 926–931.

Stengel, K., Zheng, Y., 2011. Cdc42 in oncogenic transformation, invasion, and tumorigenesis. Cell Signal 23 (9), 1415–1423.

Sternlicht, M.D., Kouros-Mehr, H., et al., 2006. Hormonal and local control of mammary branching morphogenesis. Differentiation 74 (7), 365–381.

Takahashi, S., Kubo, K., et al., 2012. Rab11 regulates exocytosis of recycling vesicles at the plasma membrane. J. Cell Sci. 125 (Pt 17), 4049–4057.

Tarone, G., Cirillo, D., et al., 1985. Rous sarcoma virus-transformed fibroblasts adhere primarily at discrete protrusions of the ventral membrane called podosomes. Exp. Cell Res. 159 (1), 141–157.

Teckchandani, A., Laszlo, G.S., et al., 2014. Cullin 5 destabilizes Cas to inhibit Src-dependent cell transformation. J. Cell Sci. 127 (Pt 3), 509–520.

Tehrani, S., Tomasevic, N., et al., 2007. Src phosphorylation of cortactin enhances actin assembly. Proc. Natl. Acad. Sci. U.S.A. 104 (29), 11933–11938.

Tepass, U., 2012. The apical polarity protein network in *Drosophila* epithelial cells: regulation of polarity, junctions, morphogenesis, cell growth, and survival. Annu. Rev. Cell Dev. Biol. 28, 655–685.

Uekita, T., Itoh, Y., et al., 2001. Cytoplasmic tail-dependent internalization of membrane-type 1 matrix metalloproteinase is important for its invasion-promoting activity. J. Cell Biol. 155 (7), 1345–1356.

Van Audenhove, I., Debeuf, N., et al., 2015. Fascin actin bundling controls podosome turnover and disassembly while cortactin is involved in podosome assembly by its SH3 domain in THP-1 macrophages and dendritic cells. Biochim. Biophys. Acta 1853 (5), 940–952.

van den Dries, K., van Helden, S.F., et al., 2012. Geometry sensing by dendritic cells dictates spatial organization and PGE(2)-induced dissolution of podosomes. Cell Mol. Life Sci. 69 (11), 1889–1901.

Varon, C., Tatin, F., et al., 2006. Transforming growth factor beta induces rosettes of podosomes in primary aortic endothelial cells. Mol. Cell Biol. 26 (9), 3582–3594.

Vogel, G.F., Klee, K.M., et al., 2015. Cargo-selective apical exocytosis in epithelial cells is conducted by Myo5B, Slp4a, Vamp7, and Syntaxin 3. J. Cell Biol. 211 (3), 587–604.

Wakabayashi, Y., Dutt, P., et al., 2005. Rab11a and myosin Vb are required for bile canalicular formation in WIF-B9 cells. Proc. Natl. Acad. Sci. U.S.A. 102 (42), 15087–15092.

Wang, J., Deretic, D., 2015. The Arf and Rab11 effector FIP3 acts synergistically with ASAP1 to direct Rabin8 in ciliary receptor targeting. J. Cell Sci. 128 (7), 1375–1385.

Way, G., Morrice, N., et al., 2002. Purification and identification of secernin, a novel cytosolic protein that regulates exocytosis in mast cells. Mol. Biol. Cell 13 (9), 3344–3354.

Wiesner, C., Faix, J., et al., 2010. KIF5B and KIF3A/KIF3B kinesins drive MT1-MMP surface exposure, CD44 shedding, and extracellular matrix degradation in primary macrophages. Blood 116 (9), 1559–1569.

Willenborg, C., Jing, J., et al., 2011. Interaction between FIP5 and SNX18 regulates epithelial lumen formation. J. Cell Biol. 195 (1), 71–86.

Williams, K.C., Coppolino, M.G., 2011. Phosphorylation of membrane type 1-matrix metalloproteinase (MT1-MMP) and its vesicle-associated membrane protein 7 (VAMP7)-dependent trafficking facilitate cell invasion and migration. J. Biol. Chem. 286 (50), 43405–43416.

Wilson, G.M., Fielding, A.B., et al., 2005. The FIP3-Rab11 protein complex regulates recycling endosome targeting to the cleavage furrow during late cytokinesis. Mol. Biol. Cell 16 (2), 849–860.

Wolf, K., Muller, R., et al., 2003. Amoeboid shape change and contact guidance: T-lymphocyte crawling through fibrillar collagen is independent of matrix remodeling by MMPs and other proteases. Blood 102 (9), 3262–3269.

Wolf, K., Wu, Y.I., et al., 2007. Multi-step pericellular proteolysis controls the transition from individual to collective cancer cell invasion. Nat. Cell Biol. 9 (8), 893–904.

Yu, S., Yehia, G., et al., 2014. Global ablation of the mouse Rab11a gene impairs early embryogenesis and matrix metalloproteinase secretion. J. Biol. Chem. 289 (46), 32030–32043.

Zhang, X.M., Ellis, S., et al., 2004. Sec15 is an effector for the Rab11 GTPase in mammalian cells. J. Biol. Chem. 279 (41), 43027–43034.

Zihni, C., Mills, C., et al., 2016. Tight junctions: from simple barriers to multifunctional molecular gates. Nat. Rev. Mol. Cell Biol. 17 (9), 564–580.

Chapter 6

Planar Cell Polarity and the Cell Biology of Nervous System Development and Disease

J. Robert Manak
The University of Iowa, Iowa City, IA, United States

INTRODUCTION

In this chapter, I will attempt to accomplish three goals: Throughout, I will summarize some of the key findings regarding the involvement of planar cell polarity (PCP) genes in development of the nervous system. Second, I will briefly discuss some of the connections between PCP genes and diseases/disorders of the nervous system. Since several fine reviews exist that comprehensively document such connections, I will only cover those germane to the topic of discussion (Goodrich, 2008; Simons and Mlodzik, 2008; Wada and Okamoto, 2009; Gray et al., 2011; Wansleeben and Meijlink, 2011; Boutin et al., 2012; Wallingford, 2012; Tissir and Goffinet, 2013; Yang and Mlodzik, 2015). Finally, and most comprehensively, I will explore some of the cell biological questions that I believe to be key in understanding PCP involvement in nervous system development. This exploration will at times be highly speculative, as there is somewhat limited data explaining precisely how PCP genes control neuronal polarization and behavior (e.g., migration, axonal advance) at the mechanistic level. The PCP field has been around for over 30 years, yet we still struggle with understanding some of the basic ground rules that underlie PCP function. This is partly due to the diverse cellular functions of these genes depending on cellular, as well as subcellular, context. In addition, many of the downstream effector genes (e.g., Rho GTPases) play more general roles in the cell, making it more difficult to identify such genes using classic mutagenic screens.

Establishing polarity is a fundamentally important developmental process, not only because it establishes front from back and top from bottom but also is ultimately critical for the physiological function of cells. For example, a migrating neuron must establish the correct polarity to ensure its proper

Cell Polarity in Development and Disease. https://doi.org/10.1016/B978-0-12-802438-6.00006-1
147

migration and to enable correct transport of signaling components along the axons (or dendrites) to the synapse, the structure that underlies electrochemical transmission from neuron to neuron, or neuron to organ. Although this chapter will mostly focus on the so-called "classic" or "core" PCP genes, their connection to apicobasal polarity (ABP) genes will also be addressed, as there is an intimate connection between these polarity axes and neuronal development.

PCP is historically defined as a process whereby cells of a tissue attain a polarity that is orthogonal to their ABP, and although early studies in *Drosophila* focused primarily on contiguous sheets of cells, mostly ectoderm, it is now clear that PCP works in a variety of tissues, including migratory cells that are not in fixed contact with their surrounding neighbors (although *are* in contact with cellular sheets and the extracellular matrix (ECM) along which they migrate). A variety of reviews detail these earlier studies in both invertebrates and vertebrates (Wada et al., 2006; Simons and Mlodzik, 2008; Bayly and Axelrod, 2011; Goodrich and Strutt, 2011; Gray et al., 2011; Wansleeben and Meijlink, 2011; Wallingford, 2012; Devenport, 2014, 2016; Sokol, 2015; Yang and Mlodzik, 2015; Butler and Wallingford, 2017). Importantly, PCP was found to not only be involved in partitioning proper structures to the correct side of the cell (such as a bristle or hair in the *Drosophila* ectoderm) but also to control the behavior of sheets of cells (i.e., vertebrate convergent extension), such that cell rearrangements create a longer, thinner sheet of cells from a shorter, wider sheet of cells.

Early genetic screens in *Drosophila* to identify loci controlling PCP identified a core set of genes (the "classic" or "core" set), which encode the membrane-associated proteins Frizzled (a seven transmembrane domain G protein-coupled receptor, which can act in some contexts as a Wnt receptor), Van Gogh (a membrane-spanning protein that is a member of the tetraspanin family), and Flamingo (a protocadherin that is a seven transmembrane domain G protein-coupled receptor), as well as the cytoplasmic proteins Prickle (a LIM domain—containing cytoplasmic protein), Dishevelled (a PDZ-containing cytosolic protein), and Diego (a cytoplasmic ankyrin repeat-containing protein) (Goodrich and Strutt, 2011; Gray et al., 2011; Yang and Mlodzik, 2015; Davey and Moens, 2017). In addition, the ligands of Frizzled, the Wnt proteins (e.g., wg, DWnt2, DWnt3/5, DWnt4, DWnt6, DWnt8, DWnt10), are used in certain contexts of PCP (but seemingly not others). In humans, two Van Gogh—like genes (VANGL1 and 2), three Flamingo genes (CELSR1—3), five Prickle-like genes (PRICKLE1—4, Testin (TES)), three Dishevelled genes (DVL1—3), and one Diego gene (Diversin or ANKRD6) exist, suggesting a strong conservation of this polarity pathway (www.genecards.org). It is important to note, however, that although I have listed paralogs that show significant homology with one another, a wide variety of genes exist which encode proteins containing conserved domains found in the core PCP proteins. One such example is LIM domain—containing proteins (of which the classic PCP protein Prickle is a member), and dozens exist (Kadrmas and Beckerle, 2004; Smith et al., 2014) including Paxillin, Zyxin, TES, and LIMK. Finally, there are up to 10 vertebrate Frizzled genes (FZD1—10) and 19 WNT genes (WNT1, 2, 2B, 3, 3A, 4, 5A, 5B, 6, 7A, 7B, 8A, 8B, 9A, 9B, 10A, 10B, 11, 16) in vertebrates, underscoring the importance of this signaling module.

Early studies in *Drosophila* showed that although initially uniformly distributed, the core proteins were found to polarize such that Van Gogh and Prickle localized to one side of an epithelial cell (proximal if considering the *Drosophila* wing), while Frizzled, Disheveled, and Diego localized to the other (distal) side (Simons and Mlodzik, 2008; Goodrich and Strutt, 2011). Each group of proteins polarized on one side of the cell was shown to interact with one other, and Flamingo was found to interact with both complexes while interacting with itself across cell boundaries, thus creating the ability to signal from one cell to another. This form of Wnt signaling is referred to as the noncanonical Wnt signaling pathway, since other studies on Wnt signaling initially revealed a signaling cascade (the canonical Wnt signaling pathway) that included both Frizzled and Disheveled, but not Van Gogh or Prickle, and this pathway was found to converge on beta-catenin stability with nuclear translocation and DNA binding, thus setting in play appropriate transcriptional programs (Nusse, 2012). Additionally, a third form of Wnt signaling was identified, namely the Wnt/calcium pathway (Thrasivoulou et al., 2013). Although there is certainly some degree of cross talk between these signaling modules, here I will mostly focus on noncanonical PCP signaling.

BREAKING THE RULES

Over the course of the early studies on PCP, a few "rules" emerged, which have more or less been debunked. First, Frizzled in the context of canonical Wnt signaling could act as a Wnt receptor, but Wnt was not necessarily needed in the context of PCP, at least in flies. However, it has now been shown that DWnt4-DFz2 is required to properly guide photoreceptor axons in *Drosophila* (Sato et al., 2006), and *Drosophila* Wnt5 was shown to regulate mushroom body (MB) axonal development along with Fz and Vang (Shimizu et al., 2011; Gombos et al., 2015). Second, Fz was thought to be the only component of the core module that could act as a Wnt receptor (which was first established in the canonical pathway and only later in vertebrate PCP). Recent reports suggest the existence of Wnt5A receptor complexes that not only includes Fz but also coreceptors Ror2 and/or Ryk and perhaps Vang (Grumolato et al., 2010; Gao et al., 2011; Andre et al., 2012). Vang and Fz might also act in separate complexes alongside Ror2 and/or Ryk to mediate Wnt5 signaling, and recent evidence provides support for this idea in fly MB Kenyon cell neurons that undergo axonal branching to form two alpha (alpha, alpha') and two beta (beta, beta') lobes. There, Vang is required for guidance and growth of beta lobes, whereas Fz is required for guidance and growth of alpha lobes (Shimizu et al., 2011; Ng, 2012; Gombos et al., 2015). Third, differential polarization of the core group of proteins was initially thought to be key to PCP, and that after polarity was established, members of the proximal complex did not associate with members of the distal complex within the same cell. Indeed, it has now been shown that some of the core PCP members comingle with members one would think should be polarized to the opposite side of the cell, such as Vangl2 and Frizzled3 (Zou, 2012). Therefore, when thinking of nervous system development, it is important to forego common PCP terms such as "proximal" and "distal," since neurons are dynamic structures that during development are constantly changing their cell shapes and orientations. Thus, a growth cone that is extending filopodia/lamellipodia (in neurons or migratory cells) should be considered a polarization center unto itself, since PCP "proximal" players such as Vangl2 or Prickle1 act within these seemingly "distal" structures (Zou, 2012; Lim et al., 2016). These ideas will be further addressed in relevant sections of the chapter.

EFFECTORS OF PLANAR CELL POLARITY GENES

A wide array of effector genes acting downstream of PCP genes have been identified. Although many were originally found in genetic studies of PCP in the fly eye, wing, and notum, I will focus mainly on effectors that are also known to act in nervous system development. These include the formin dDAAM (Habas et al., 2001; Matusek et al., 2006), dRac1 (Eaton et al., 1995; Fanto et al., 2000; Habas et al., 2003; Munoz-Descalzo et al., 2007), dRhoA (Strutt et al., 1997; Fanto et al., 2000; Habas et al., 2001; Tahinci and Symes, 2003; Munoz-Descalzo et al., 2007), and Rho Kinase (drok) (Winter et al., 2001; Marlow et al., 2002), and the formin-like multiple wing hairs (Lu et al., 2015). RhoA and Rac1, as well as Cdc42, belong to the Rho family of small GTPases that are heavily involved in modulating cytoskeletal dynamics, and all play a role in axonal extension dynamics in neurons.

AXONAL DETERMINATION/POLARIZATION: CONNECTIONS TO POLARITY GENES

Two key goals of establishing polarity are (1) to create an organization of cytoskeletal components such that cellular components of a cell can be transported to their proper positions (which can help amplify polarity) and (2) to create a dynamic cytoskeletal system that can read cues from its environment to respond and reorganize so that cells or cellular processes can move in a directed fashion. The latter process uses cellular structures such as lamellipodia/filopodia, which are observed at the growth cone of migrating cells (including neurons) as well as advancing axons of neurons whose cell bodies are stationary. Achieving both of these goals requires proper establishment, or refinement in the case of migrating cells, of both microtubules (MTs) and actomyosin cytoskeletal elements. Organization of both types of cytoskeletal elements must be intimately coordinated, and current research is focused on understanding how these cytoskeletal elements talk to one another at the molecular level.

Extension of a growth cone from an axon not only requires modulation of cytoskeletal dynamics but also constant assembly and turnover of cellular adhesions in the region of the growth cone (Short et al., 2016). These adhesions are oftentimes referred to as point contacts (PCs) (called focal adhesions (FAs) in other cell types) and help provide the anchor points for growth cone advance. Additionally, PCs and the cytoskeleton engage in constant cross talk and can influence the dynamics of each other as will be discussed below. The growth cone itself is made up of both lamellipodia and filopodia. The lamellipodia are primarily involved in movement of the growth cone, whereas filopodia are involved with sensing the environment for guidance cues (Dent et al., 2011). The growth cone looks somewhat like a duck's webbed foot, with the webs being lamellipodia and the "toes" being the filopodia. Below I will discuss two of the key aspects of development of a neuron: First, I will describe axonal determination, followed by axonal extension (or advance, with or without migration of the neuron as a whole). All neurons must elaborate an axon or axon-like structure, but only some neurons will migrate. Axonal extension involves pathfinding by growth cones at the tip of the axon, which sense guidance factors in the ECM (Myers et al., 2011).

Before a neuron becomes polarized and forms an axis-defining axon, it is essentially a cell body extending neurites in random directions to probe for instructive information (Takano et al., 2015). Eventually, based on the presence of local signaling molecules present in the ECM, one neurite begins the process of becoming an axon, and this is accompanied by enhancement of the MTs cytoskeleton such that robust MT growth increases the axon length, as well as an increase in the dynamics of the actin cytoskeleton. These processes are under the control of the Rho GTPases, and MT-based vesicle trafficking is increased to provide membrane material for growth of the axon (Arimura and Kaibuchi, 2007) as well as potentially amplify the transport of polarity proteins such as Vang and the Par polarity complex (see below). However, it should be noted that local signaling molecules can influence growth of dendrites as well (Whitford et al., 2002). RhoA is a negative regulator of neurite formation and appears to be expressed more highly in growth cones of minor neurites *not* destined to become the axon (Da Silva et al., 2003; Pertz, 2010). However, RhoA can also stimulate polymerization of actin through formins to produce actin filaments (Carlier et al., 2015). Formins possess MT-binding activity to modulate both MT and actin dynamics (Spiering and Hodgson, 2011), thereby providing a link between these two cytoskeletal elements. Rho-dependent actin fibers are stabilized in the minor neurites by myosin II and profilin IIa, preventing the MTs from penetrating them (Da Silva et al., 2003; Kollins et al., 2009). The combination of actin fibers and myosin II thus produce so-called "stress fibers," which are able to generate the force necessary for retraction of the minor neurites (Katoh et al., 2011). Interestingly, activation of RhoA has been shown to be mediated by Dishevelled during Wnt3a-induced retraction of neurites in two different neuronal cell lines (Kishida et al., 2004; Tsuji et al., 2010). The theme of RhoA control of minor neurite retraction will emerge once again in migrating neurons, described later.

The balance of MTs versus actin filaments helps determine which neurite becomes an axon; in such a case, MTs are stabilized through associated proteins such as Tau, protecting them from severing enzymes (Witte and Bradke, 2008). At the same time, actin filaments are destabilized in the growing axon, ultimately leading to deeper penetration of MTs and thus axon specification (Bradke and Dotti, 1999; Flynn et al., 2012). Cheng et al. (2011) have shown that phosphorylation of a ubiquitin E3 ligase (Smurf1) in response to extracellular brain-derived neurotrophic factor signaling in the neurite destined to become the axon alters Smurf1 preference for its targets, thus enhancing the stability of a polarity-promoting factor (Par6), while at the same time promoting degradation of the growth-inhibiting RhoA. This may further stabilize the Par3/Par6/aPKC complex (a complex known to regulate ABP in flies; (Chen and Zhang, 2013)) in the axon, which recruits the active form of Cdc42 to modulate actin reorganization. It is worth noting that the aforementioned Par complex is localized at the growth cone (Shi et al., 2004; Insolera et al., 2011), at the distal end of a growing axon, and thus defines a novel location for this complex for those used to thinking about Par complex localization at adherens junctions (AJs) in the fly ectoderm. Overexpression of Par3 can induce multiple axons, as can overexpression of activated forms of polarity effectors Cdc42 (Schwamborn and Puschel, 2004) or Rap1 (the latter of which appears to localize as well as activate the Par complex (Schwamborn and Puschel, 2004; Gerard et al., 2007)). Also, Shootin1 can increase anterograde transport in the axon and thus help promote axon growth and neuronal polarization (Toriyama et al., 2006). Increased anterograde transport would also be promoted by an increased number of plus-end-out MTs during polarization, and this has been documented in several studies (Yu and Baas, 1994; Seetapun and Odde, 2010; Kapitein and Hoogenraad, 2011; Yau et al., 2016). *Drosophila* Prickle has been shown to modulate MT polarity (including in neurons) but also affect vesicular transport along MTs (Ehaideb et al., 2014; Olofsson et al., 2014), and a reduction in the levels of anterograde Kinesin motor proteins suppresses *prickle*-mediated seizures (Ehaideb et al., 2014). These data suggest a potential role for Prickle in axonal determination, although this has yet to be determined.

PLANAR CELL POLARITY MUTANTS EXHIBIT AXONAL EXTENSION DEFECTS

Upon neuronal polarization, axonal advance/extension is necessary for responding to pathfinding cues in the environment and moving the axon toward a target. Studies in both invertebrates and vertebrates have documented axonal extension defects in subsets of neurons in animals mutant for a variety of PCP genes. In *Drosophila*, abdominal sensory neuron axons failed to reach their target in loss-of-function mutants of the *pk^{pk}* isoform of *prickle* (Mrkusich et al., 2011), whereas axon extension and branching defects were observed in Kenyon cell neurons of the fly MB in a variety of PCP mutants, including *pk, fz, stbm, dsh*, and *fmi* (Shimizu et al., 2011; Ng, 2012; Gombos et al., 2015). One of these studies provides strong evidence directly connecting several core PCP members with two of the aforementioned PCP effectors known to be involved in neuronal development (Gombos et al., 2015). There, a signaling cascade was identified that includes Wnt5, Frizzled, Strabismus, and Dishevelled that act together with PCP effectors Rac1 and dDAAM to regulate actin cytoskeleton dynamics of the axonal growth cone, thereby promoting axonal branching and extension (Gombos et al., 2015). Interestingly, loss of the *prickle spiny-legs* isoform (*pk^{sple}*) yields MB defects, but only rarely if at all are defects seen in *pk^{pk}* mutants (Shimizu et al., 2011; Gombos et al., 2015) (unpublished data), consistent with the hypothesis that the different *prickle* isoforms may have independent functions in different sets of neurons. In the ectoderm, this appears to be the case in that Pk^{pk} is required only in a subset of ectodermal cells, whereas Pk^{sple} is required in a complementary set of cells to rectify inverted upstream information gradients that would be predicted to invert polarity of bristles and hairs (Ayukawa et al., 2014; Olofsson et al., 2014). Alternatively, given previous observations that both *pk* isoforms are coexpressed in ectodermal tissues (albeit with elevated expression of one isoform over the other depending on context; (Ayukawa et al., 2014; Olofsson et al., 2014)), it is possible that both isoforms are needed for independent functions in cells. Consistent with this idea, we find that both *pk* isoforms appear to be expressed at comparable levels in commissural neurons in the ventral nerve cord of third instar *Drosophila* larvae (Lilienthal and Manak, unpublished data). If this is true, it follows that the differences in amino acid sequence of the two isoforms should dictate the different functions in neurons. The Pk^{pk} isoform has 13 unique N terminal amino acids not found in Pk^{sple}, whereas the Pk^{sple} isoform has 349 N terminal amino acids not found in Pk^{pk}. Intriguingly, using the Eukaryotic Linear Motif resource for functional sites in proteins (ELM; http://elm.eu.org/), we find the unique N terminal domain of Pk^{sple} contains several short proline-rich peptide sequences recognized by class I (both canonical and noncanonical) and class II SH3 domains that are also found in MT binding, AJ, and FA proteins, including TAU, FAK1,

PAK1, WASF1, and DNM1 (unpublished data). Although the relevance of these domains needs to be confirmed, these data suggest the intriguing possibility that Pk^{sple} may be targeted to FA-like PCs in neurons, which would align nicely to the recent observation that Prickle1 associates with FAs in gastric cancer cells (Lim et al., 2016).

Elegant work by the Zou group has shown that in vertebrates, axonal pathfinding in commissural neurons of the neural tube is controlled (in part) by Wnt5a and PCP components (Lyuksyutova et al., 2003; Shafer et al., 2011; Onishi et al., 2013). Notably, Vangl2 becomes enriched in growth cone filopodia nearest the source of Wnt5a, and this leads to Fzd3 internalization, which is dependent on the endocytic recycling GTPase Arf6 (Shafer et al., 2011; Onishi et al., 2013). This work demonstrates "breaking" the rule of asymmetry by the core PCP components, in this case by Vangl2 and Fzd3 occupying the same relative domain of the cell membrane. How internalization of Fzd3 vesicles is accomplished is currently unknown; however, work from *Drosophila* may again provide some clues. There, Van Gogh and Prickle not only help establish the proper polarity of MTs, but also affect the vesicle transport on them (Ehaideb et al., 2014; Olofsson et al., 2014). Thus, in vertebrate growth cone filopodia, localization of Vangl2, perhaps with its known association partner Prickle, may help regulate not only organization of MTs within the growth cone but also the transport of endocytosed Fzd-containing vesicles internally toward the cell body, reducing the chances of the vesicle from returning to, and fusing with, the membrane. Indeed, the work in gastric cancer cells has shown that Prickle1 not only associates with FAs but is also in complex with plus-end MT-binding proteins (CLASP1 and 2) to tether the MTs to the adhesions (Lim et al., 2016). It is thus theoretically possible that Prickle could be serving a similar function at the growth cone with its classical binding partner, Vangl, in this case using CLASPs to capture MTs at the leading edge. A speculative model is shown in Fig. 6.1.

AXONAL EXTENSION AND THE DYNAMICS OF THE CYTOSKELETON

Coordination and dynamic regulation of both the MT and actin cytoskeletons are critical in moving the axonal growth cone. The FA-like PCs are necessary at the growth cone for anchoring it to the ECM, but also for capturing and holding actin fibers via cell adhesion molecules. The clutch hypothesis provides a mechanism for how growth is accomplished (Mitchison and Kirschner, 1988). First, actin filaments are anchored to the adhesion sites, preventing retrograde flow of actin (the default state), and this is followed by an increase in actin filament dynamics whereby G-actin is added to the barbed end of F-actin fibers thus creating protrusive activity at the growth cone, while MTs continue to grow into more peripheral regions (Suter et al.,

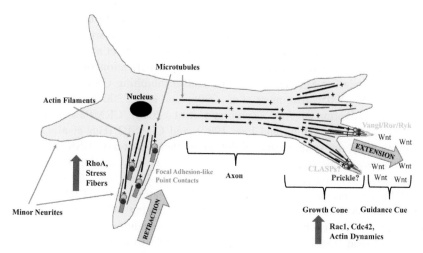

FIGURE 6.1 Model for Prickle function in neurons. This model imagines at least two neuronal functions for Prickle: First, Prickle (*purple circles*) is working with CLASP proteins (*orange circles*) in minor neurites to capture and tether plus-end out microtubules (MTs) to the point contacts (*blue boxes*), which destabilize them. Stress fibers (actin fibers shown in red, myosin II not shown) can then retract the neurites. The minor neurites are elevated for RhoA activity. Second, Prickle is working during axonal pathfinding with Vangl and coreceptors such as Ror or Ryk (*green outlines*) to transduce Wnt guidance cues. This could be accomplished by Prickle (bound to Vangl) and CLASPs capturing the plus-ends of MTs in much the same manner as seen at focal adhesions, which ultimately results in protrusive force generation assisted by an increase in actin dynamics (see text for details). The growth cone has elevated levels of Rac1 and Cdc42.

1998; Bard et al., 2008; Schaefer et al., 2008; Hyland et al., 2014; Garcia et al., 2015), potentially contributing to the force generation (Lu et al., 2013; Roossien et al., 2013). To accomplish the actin-based protrusion, Cdc42 activates cofilin, an actin-binding protein that disassembles actin filaments to produce G-actin (Garvalov et al., 2007). The resulting G-actin monomers are then free to be added onto the barbed end of the anchored F-actin filaments at the growth cone, thus promoting protrusive forces (Carlier et al., 2015). Although retrograde flow of actin is prevented, Myosin II nonetheless creates a retrograde force that aids in growth cone advance (Mitchison and Kirschner, 1988). Notably, actin polymerization and protrusive force is upregulated by the PCP effector and GTPase Rac1 (Tahirovic et al., 2010). Rac1 acts by mobilizing both the WAVE (WASP [Wiskott—Aldrich syndrome proteins]-family verprolin-homologous protein) and Arp2/3 proteins to promote branching of actin filaments thus leading to membrane protrusion at the edge of the lamellipodium/growth cone (Takenawa and Suetsugu, 2007; Schelski and Bradke, 2017).

MTs are also thought to play a role in generating protrusive force in the growth cone. Here, I will summarize some of the key findings in this area; for a more detailed summary, an excellent review has recently been published (Schelski and Bradke, 2017). In the growth cone, MTs are stabilized by Rac1-mediated inactivation of stathmin (Watabe-Uchida et al., 2006), which normally either depolymerizes or prevents polymerization of MTs. PI3K, another important MT regulator, is critical in inhibiting GSK3beta activity (Jiang et al., 2005; Yoshimura et al., 2006; Guo et al., 2007), which in turn leads to activation of APC (Zumbrunn et al., 2001), CRMP2 (Yoshimura et al., 2006), Tau (Hong and Lee, 1997), and CLASP2 (Hur et al., 2011). Additionally, in spite of its inhibition in these contexts, GSK3beta can nonetheless phosphorylate MAP1b (Trivedi et al., 2005), leading to increased MT polymerization. APC, CRMP2, Tau, and CLASP2 are MT-binding proteins with roles in MT stabilization (APC, CRMP2, Tau, CLASP2), MT assembly (CRMP2), and/or promoting MT-mediated vesicle transport (APC) (Schelski and Bradke, 2017). In addition, CLASP proteins, MT plus-end tracking proteins (+TIPs), are able to tether MTs to FAs to stimulate their disassembly (Stehbens et al., 2014), an observation that will become important later when discussing Prickle function. Finally, growth cone–turning requires both Wnt5a as well as Tau (Li et al., 2014), directly tying MT structural organization with PCP pathway–induced growth cone behavior.

LAMELLIPODIA AND FILOPODIA USE DIFFERENT ACTIN NUCLEATORS

As mentioned above, filopodia primarily act as the environmental sensors, whereas lamellipodia are involved in movement of the growth cone. Both structures require actin filament assembly, but whereas filopodia contain unbranched actin fibers tethered together by Fascin and anchored to the membrane by Formins, lamellipodia contain actin filaments that are nucleated into branched structures by the SCAR/WAVE proteins (Pollard and Borisy, 2003) in concert with Arp2/3, the protein that tethers the fibers at the branch points (Carlier et al., 2015). SCAR/WAVE proteins fall under the family of WASP proteins (Carlier et al., 2015). Additionally, Ena/VASP proteins are involved in actin filament elongation in filopodia/lamellipodia (Breitsprecher et al., 2011). When Mammalian ENA (Mena)/VASP/EVL (all Ena/VASP proteins) are triply mutated, mice show defects in axon formation in the cortex likely due to a failure to form filopodia (Kwiatkowski et al., 2007). Moreover, the formin mDia interacts with Ena/VASP proteins to modulate actin dynamics (Grosse et al., 2003; Shakir et al., 2006; Fleming et al., 2010). Intriguingly, TES, the protein most similar to the Prickles (all containing one PET and three LIM domains) not only genetically and physically interacts with Vangl2 in the mouse to regulate inner ear sensory cell orientation, but binds to Mena (Boeda et al., 2007), connecting this Prickle-like protein with core PCP members and the actin cytoskeleton.

NEURONAL CELL MIGRATION USES THE SAME PLAYERS AS DURING AXONAL POLARIZATION

Analogous to axonal determination where a single neurite becomes the axon and minor neurites are retracted, a migrating neuron uses the same signaling components to accomplish its goals. For example, at the leading edge, Cdc42 and Rac1 GTPases drive growth cone advance through modulation of the actin and MT cytoskeletons. At the trailing edge, the RhoA GTPase is not only utilized to destabilize any competitor processes which might form, but also to promote disassembly of PCs at the rear of the cell. However, an added feature at play during neuronal migration is the involvement of the MT cytoskeleton in nuclear translocation. In addition to providing scaffolding in the axon, MTs also form a cage-like structure around the nucleus (Rivas and Hatten, 1995; Moon and Wynshaw-Boris, 2013), and its organization is mediated by the focal adhesion kinase, FAK (Xie et al., 2003). Pulling force mediated by the cytoskeleton can then translocate the nucleus closer to the growth cone, and this process involves both Lis1 and Dynein, both of which are part of the minus-end directed MT motor complex (Tsai et al., 2007). Therefore, whether a neuron disassembles PCs before retracting minor neurites during axon specification or disassembles PCs in regions other than the growth cone leading edge during axonal extension/neuronal migration, RhoA is a key modulator.

Several PCP mutants reveal neuronal migration defects. The Prickle protein Pk1a is required in the neuroepithelium along which facial branchiomotor neurons (FBMNs) migrate, as are other PCP proteins such as Vangl2, Fz3a, Stbm, and several Celsr paralogs (Jessen et al., 2002; Formstone and Mason, 2005; Wada et al., 2005, 2006; Walsh et al., 2011), while Pk1b is required cell-autonomously in the FBMN neurons themselves (Carreira-Barbosa et al., 2003) along with other polarity proteins such as Scrib and Vangl2 (Walsh et al., 2011). Further work by the Moens group has recently shown that Fzd3a and Vangl2 act to stabilize or destabilize growth cone filopodia, respectively, in the FBMNs (Davey et al., 2016). Strikingly, the opposite is true for the neuroepithelium; there, Vangl2 is required for FBMN filopodial stabilization, and Fzd3a for filopodial destabilization. The authors propose an intriguing model that relies on the mechanism of classical PCP in fly ectoderm; namely, that Van Gogh complexes specifically associate with Fzd complexes across membranes. Thus, a Fzd3a-presenting FBMN filopodium in exploratory mode would preferentially interact with a neuroepithelial cell presenting its complementary partner, Vangl2. As the authors point out, alternative models may also be at play, such as PCP-dependent trafficking of cadherins or a PCP-driven reduction of ECM (Williams et al., 2012; Warrington et al., 2013; Nagaoka et al., 2014). Finally, a recent study identified a *PRICKLE1* point mutation in a LIM domain as causing a smooth brain phenotype in humans, reminiscent of the lissencephaly phenotype caused by defective neuronal migration during neocortex development (Bassuk and Sherr, 2015).

THE ADHERENS JUNCTION: A CONVERGENCE OF INTERDEPENDENT SIGNALING COMPLEXES

In the epidermis of the fly, the PCP complexes localize at the level of AJs (also known as cadherin junctions), precisely where proteins controlling ABP localize, including cdc42, aPKC, Pars 3 and 6, and Crumbs. The AJ is used to physically connect one cell to another via the cytoskeleton, and it is important to note that AJs show many similarities to FAs, which adhere cells to the cell matrix (Padmanabhan et al., 2015). Additionally, there is ample cross talk between the various complexes of the ABP machinery and the PCP machinery (Elsum et al., 2012), in some cases through AJs, and other cases through FAs. Although developing neurons do not contain AJs analogous to those that exist for epithelial cell sheets, they do contain FA-like PCs, and as discussed these are critical for extension of neurites or advance of axons as well as migration of neurons from one location to another (Short et al., 2016). Notably, Prickle 1 was recently shown to be localized adjacent to FAs and is required to promote disassembly of these structures during cell migration of gastric cancer cells (Lim et al., 2016).

The AJ (which in neurons is a constituent of synaptic junctions) is made up of many proteins, including various membrane-spanning cadherins, p120 plakoglobin, and alpha- and beta-catenin (see Table 6.1). These proteins, many of which are conserved down to unicellular organisms, form a complex across cell membranes, and are connected to the actin cytoskeleton via beta-catenin and plakaglobin (Harris and Tepass, 2010; Brieher and Yap, 2013; Zaidel-Bar, 2013; Padmanabhan et al., 2015). Additionally, a large number of ancillary proteins and complexes reside at the AJ. A primary function for this complex is to connect the cytoskeleton of one cell to the next, transmitting tensile forces from outside to within. Of note, stretching of the AJ induces conformational changes in protein components such as Vinculin, which likely leads to recruitment of Arp2/3 and VASP (Padmanabhan et al., 2015). As previously discussed, Ena/VASP proteins play critical roles in bundling f-actin fibers in neurons, which provide force against the cell membrane in order to form the filopodia required for axonal extension/neuronal migration (Kwiatkowski et al., 2007). In addition to the classical AJ and PCP complexes of proteins, several additional complexes exist at or near this subcellular structure, namely the Crumbs/Stardust, PATJ/Lin-7 complex (Bulgakova and Knust, 2009), the Par3/Par6/aPKC complex (Chen and Zhang, 2013), and the more basally localized Scribble/Lethal giant larvae/ Discs large complex (Su et al., 2012). Evidence has been accumulating for interactions across complexes; for example, Par3/Par6/aPKC can bind Lgl and inactivate it in neurons (Betschinger et al., 2003), and a genetic

TABLE 6.1 Comparison of the Adherens Junction and Focal Adhesion Components

Adherens Junction	Protein Name	Category	Focal Adhesion	Protein Name	Category
CTNNA1	catenin alpha 1	actin-binding adaptor	CORO1B	coronin 1B	actin regulation
CTNNA2	catenin alpha 2	actin-binding adaptor	CORO2A	coronin-2A	actin regulation
CTNNA3	catenin alpha 3	actin-binding adaptor	NEXN	nelin	actin regulation
DBN1	drebrin 1	actin-binding adaptor	PFN1	profilin	actin regulation
FSCN1	fascin homolog 1, actin-bundling protein	actin-binding adaptor	GRB2	grb2	adaptor
LIMA1	LIM domain and actin-binding protein 1	actin-binding adaptor	LDB3	zasp	adaptor
SHROOM3	shroom family member 3	actin-binding adaptor	LIMS1	PINCH1	adaptor
ANK3	ankyrin-G	adaptor	LIMS2	PINCH2	adaptor
CTNNB1	catenin beta 1	adaptor	TNS1	tensin	adaptor
CTNND1	catenin delta 1	adaptor	FHL2	FHL2	adaptor
CTNND2	catenin delta 2	adaptor	LPXN	leupaxin	adaptor

Continued

TABLE 6.1 Comparison of the Adherens Junction and Focal Adhesion Components—cont'd

Adherens Junction	Protein Name	Category	Focal Adhesion	Protein Name	Category
DLG1	disks large homolog 1	adaptor	TNS2	tensin 2	adaptor
DLG5	disks large homolog 5	adaptor	ITGA1 (integrin)	CD49a	adhesion receptor
JUP	catenin gamma	adaptor	ITGA2	CD49b	adhesion receptor
LMO7	LIM domain only protein 7	adaptor	ITGA4	CD49d	adhesion receptor
NUMB	protein numb homolog	adaptor	ITGA5	CD49e	adhesion receptor
PKP4	plakophilin 4	adaptor	ITGA7	ITGA7	adhesion receptor
SCRIB	scribbled homolog	adaptor	ITGA8	ITGA8	adhesion receptor
PSEN1	presenilin 1	arotease	ITGA9	ITGA9	adhesion receptor
DIAPH1	diaphanous related formin 1	actin dynamics regulator	ITGA10	ITGA10	adhesion receptor
WASF2	wiskott–Aldrich syndrome protein family member 2	actin dynamics regulator	ITGA11	ITGA11	adhesion receptor
WASL	neural Wiskott–Aldrich syndrome protein	actin dynamics regulator	ITGAD	CD11D	adhesion receptor
CDH1	cadherin 1	adhesion receptor	ITGAE	CD103	adhesion receptor
CDH10	cadherin 10	adhesion receptor	ITGAL	CD11a	adhesion receptor

CDH11	cadherin 11	adhesion receptor	ITGAM	CD11b	adhesion receptor
CDH12	cadherin 12	adhesion receptor	ITGAV	CD51	adhesion receptor
CDH15	cadherin 15	adhesion receptor	ITGAW	ITGAW	adhesion receptor
CDH18	cadherin 18	adhesion receptor	ITGAX	CD11c	adhesion receptor
CDH19	cadherin 19	adhesion receptor	ITGB2	CD18	adhesion receptor
CDH2	cadherin 2	adhesion receptor	ITGB3	CD61	adhesion receptor
CDH20	cadherin 20	adhesion receptor	ITGB4	CD104	adhesion receptor
CDH22	cadherin 22	adhesion receptor	ITGB5	ITGB5	adhesion receptor
CDH24	cadherin 24	adhesion receptor	ITGB6	ITGB6	adhesion receptor
CDH3	cadherin 3	adhesion receptor	ITGB7	ITGB7	adhesion receptor
CDH4	cadherin 4	adhesion receptor	ITGB8	ITGB8	adhesion receptor
CDH5	cadherin 5	adhesion receptor	LRP1	LRP-1	adhesion receptor
CDH6	cadherin 6	adhesion receptor	DOCK1	DOCK1	GEF
CDH7	cadherin 7	adhesion receptor	ELMO1	ELMO	GEF
CDH8	cadherin 8	adhesion receptor	DNM2	dynamin	GTPase
CDH9	cadherin 9	adhesion receptor	LIMK1	LIMK	serine/threonine kinase
FAT1	FAT atypical cadherin 1	adhesion receptor	PRKCA	PKC	serine/threonine kinase
EXOC3	exocyst complex component 3	trafficking regulator	TRPM7	trpm7	channel

Continued

TABLE 6.1 Comparison of the Adherens Junction and Focal Adhesion Components—cont'd

Adherens Junction	Protein Name	Category	Focal Adhesion	Protein Name	Category
ITGA3	CD49c	adhesion receptor	ITGA3	CD49c	adhesion receptor
ITGA6	CD49f	adhesion receptor	ITGA6	CD49f	adhesion receptor
ITGB1	CD29	adhesion receptor	ITGB1	CD29	adhesion receptor
PXN	paxillin	adaptor	PXN	paxillin	adaptor
SDCBP	syntenin 1	adaptor	SDCBP	syntenin-1	adaptor
CRKL	CrkL	adaptor	CRKL	CrkL	adaptor
TES	TES	adaptor	TES	TES	adaptor
ACTN1	alpha-actinin 1	actin-binding adaptor	ACTN1	alpha-actinin-1	actin-binding adaptor
EZR	ezrin	actin-binding adaptor	EZR	ezrin	actin-binding adaptor
KEAP1	kelch-like ECH-associated protein 1	actin-binding adaptor	KEAP1	kelch-like ECH-associated protein 1	actin-binding adaptor
MSN	moesin	actin-binding adaptor	MSN	moesin	actin-binding adaptor
RDX	radixin	actin-binding adaptor	RDX	radixin	actin-binding adaptor
SVIL	supervillin	actin-binding adaptor	SVIL	supervillin	actin-binding adaptor

TLN1	talin 1	actin-binding adaptor
VCL	vinculin	actin-binding adaptor
FBLIM1	filamin-binding LIM protein 1	adaptor
ILK	Integrin-linked protein kinase	adaptor
JUB	ajuba	adaptor
LPP	lipoma-preferred partner	adaptor
SORBS3	vinexin	adaptor
TRIP6	thyroid receptor–interacting protein 6	adaptor
ZYX	zyxin	adaptor
CTTN	src substrate cortactin	actin dynamics regulator
ENAH	ENAH, actin regulator	actin dynamics regulator
VASP	vasodilator-stimulated phosphoprotein	actin dynamics regulator
PKD1	polycystin 1	transmembrane
ACTB	actin	cytoskeleton

Continued

TABLE 6.1 Comparison of the Adherens Junction and Focal Adhesion Components—cont'd

Adherens Junction	Protein Name	Category	Focal Adhesion	Protein Name	Category
VIM	vimentin	cytoskeleton	VIM	vimentin	cytoskeleton
MYH9	myosin, heavy chain 9, nonmuscle	motor	MYH9	myosin, heavy chain 9, nonmuscle	motor
RAC1	ras-related C3 botulinum toxin substrate 1	GTPase	RAC1	ras-related C3 botulinum toxin substrate 1	GTPase
RHOA	ras homolog gene family, member A	GTPase	RHOA	ras homolog gene family, member A	GTPase
DLC1	Rho GTPase-activating protein 7	RhoGAP	DLC1	Rho GTPase-activating protein 7	RhoGAP
TIAM1	T-lymphoma invasion and metastasis-inducing protein 1	RhoGEF	TIAM1	T-lymphoma invasion and metastasis-inducing protein 1	RhoGEF
PTEN	phosphatidylinositol-3,4,5-trisphosphate 3-phosphatase and dual-specificity protein phosphatase PTEN	phosphatidylinositol phosphatase	PTEN	phosphatidylinositol-3,4,5-trisphosphate 3-phosphatase and dual-specificity protein phosphatase PTEN	phosphatidylinositol phosphatase
MAPK8	JNK	serine/threonine kinase	MAPK8	JNK	serine/threonine kinase

PAK1	serine/threonine-protein kinase PAK 1	serine/threonine kinase	PAK1	serine/threonine-protein kinase PAK 1	serine/threonine kinase
ROCK1	Rho-associated, coiled-coil containing protein kinase 1	serine/threonine kinase	ROCK1	Rho-associated, coiled-coil containing protein kinase 1	serine/threonine kinase
PTK2B	protein-tyrosine kinase 2-beta	tyrosine kinase	PTK2B	protein-tyrosine kinase 2-beta	tyrosine kinase
SRC	protooncogene tyrosine-protein kinase Src	tyrosine kinase	SRC	protooncogene tyrosine-protein kinase Src	tyrosine kinase
CAPN1	calpain-1 catalytic subunit	protease	CAPN1	calpain-1 catalytic subunit	protease
FLNA	filamin	actin regulation	FLNA	filamin	actin regulation
ACTR3	arp3	actin regulation	ARPC2	arp2/3	actin regulation

The first/second and fourth/fifth columns list many of the key protein subunits of the adherens junction and focal adhesion, respectively. Components unique to one or the other are shaded in different hues of beige, and shared components are shaded in yellow. Different components of the same complex boxed in blue. The third and sixth columns list the function of the protein subunits (Category).
Modified from Harris, T.J., Tepass, U., 2010. Adherens junctions: from molecules to morphogenesis. Nat. Rev. Mol. Cell Biol. 11, 502–514; Zaidel-Bar, R., 2013. Cadherin adhesome at a glance. J. Cell Sci. 126, 373–378; and Padmanabhan, A., Rao, M.V., Wu, Y., Zaidel-Bar, R., 2015. Jack of all trades: functional modularity in the adherens junction. Curr. Opin. Cell Biol. 36, 32–40.

interaction was discovered in mice between Vangl2 and Scrb1 with regard to polarization of stereociliary bundles in the mouse cochlea (Montcouquiol et al., 2003). Subsequently, Scrb1 was shown to directly interact with Vangl2 through its multiple PDZ domains (Kallay et al., 2006). Additional work has shown that Fz1 can bind dPatj, and mutants of either produce PCP defects in the fly eye (Djiane et al., 2005). Regarding nervous system development, Scribble1 has been implicated in neural migration of FBMNs in zebrafish embryos (Wada et al., 2005), similar to what has been observed for PCP genes (Jessen et al., 2002; Carreira-Barbosa et al., 2003; Formstone and Mason, 2005; Wada et al., 2006). In this study, it was shown that zebrafish scrb1 was required non-cell autonomously, showing robust expression in the dorsal neural tube cells that surround the FBMNs. Finally, Par6 and Par3 form a complex with aPKC, which leads to axon formation in neurons through Rac1 activation (Nishimura et al., 2005; Arimura and Kaibuchi, 2007). Thus, several lines of evidence are pointing toward interactions between AJ and polarity proteins.

FOCAL ADHESIONS: REINVENTING THE ADHERENS JUNCTION FOR CELLULAR MOVEMENT AND MIGRATION

Similar to AJs, FAs also link the actin cytoskeleton with transmembrane proteins (integrins); however, rather than connecting neighboring cells, these transmembrane proteins interact with the ECM (Winograd-Katz et al., 2014). AJs utilize cadherins and catenins to connect one cell with another, and FAs utilize integrins to connect cells to the ECM. Remarkably, however, AJs and FAs share many of the same proteins (see Table 6.1), including adaptors Paxillin, Zyxin, alpha Actinin, Ezrin, Vinculin, and Talin; GTPases RhoA and Rac1; kinases PAK1, SRC, and ROCK1; RhoGAPs/GEFs DLC1 and TIAM1; and actin dynamics regulators ENAH (enabled), VASP, and Cortactin, amongst several other proteins; excellent reviews are available that catalog components of these complexes (Harris and Tepass, 2010; Zaidel-Bar, 2013; Winograd-Katz et al., 2014) and highlight the overlap (Padmanabhan et al., 2015). Of particular note, several components of AJs and FAs are LIM domain–containing proteins such as TES, Paxillin, FBLIM1, Ajuba, LPP, TRIP6, and Zyxin (all found in both), LIMA1 and LMO7 (found only in AJs), and LASP1, FHL2, LIMS1, LIMS2, LDB3, and LIMK1 (found only in FAs). TES acts as a tumor suppressor, and its overexpression suppresses cell growth (Boeda et al., 2011). It associates with Zyxin through its first LIM domain, and the third LIM domain can displace Mena (required for actin polymerization) away from the leading edge when overexpressed. It has been suggested that loss of TES favors Mena-dependent changes in cell migration/adhesion, leading to invasive migration (Boeda et al., 2011). TES plays a role in the regulation of the actin cytoskeleton, and its knockdown in

HeLa cells reduces RhoA activity as well as the number of actin stress fibers (Griffith et al., 2005). As has been discussed, TES is the AJ/FA protein most similar to the Prickle proteins, which have a one PET-three LIM domain architecture. Notably, few LIM domain—containing proteins have been shown to also contain PET domains (for example, the multiple Prickles, Espinas, TES), suggesting a more intimate connection between TES and Prickle than between TES other LIM domain containing proteins. Given that so many PCP proteins interact and work together with apicobasal proteins at the AJ, it is likely that similar roles will continue to be discovered at the FA.

In flies, *prickle* mutants show axonal extension defects in MB neurons as well as abdominal sensory neurons (Mrkusich et al., 2011; Shimizu et al., 2011; Ng, 2012). Additionally, a recent study has identified a role for Prickle1 in promoting FA turnover (a process required for neuronal migration as indicated above) in MKN1 gastric cancer cells by localizing to cell projection-like structures targeted for retraction at the cell periphery, thus only allowing a single protrusion to form in the correct plane of migration (Lim et al., 2016). There, Prickle1 was found in close proximity with Paxillin, adjacent to FAs. Moreover, studies in migratory breast cancer cells demonstrated a role for Pk1 and Arhgap21/23 in suppressing cellular protrusions from membranes lateral to the active protrusions directing migration (Zhang et al., 2016). These data fit well with results obtained for the zebrafish *prickle1b* morphants, which suggest that *prickle1b*-deleted FBMNs fail to retract exploratory protrusions extending in directions other than the intended caudal tangential migratory path (Mapp et al., 2010). These data present an intriguing perspective on neuronal migration (and axon advance as well), suggesting that directed migration/axonal extension may be controlled by eliminating competitor processes rather than simply extending a single process in the correct plane of migration. It is important to note that growing MTs that extend into FAs promote their disassembly (Akhshi et al., 2014), and as mentioned above, Prickle plays important roles in modulating MT polarity and dynamics (Ehaideb et al., 2014; Olofsson et al., 2014). In addition, Prickle1 associates with the MT plus-end tracking proteins CLASP1 and CLASP2 (Lim et al., 2016), which are able to tether MTs to FAs to stimulate their disassembly (Stehbens et al., 2014). Coalescing these data together with the *Prickle* mutant phenotypes in neurons and cancer cells suggests a possible model whereby Prickle in concert with CLASPs might be guiding or directing MT plus-ends toward the FA-like PCs in neurites requiring retraction, and this is followed by PC disassembly and stress fiber activation to retract these neurites. Thus, a Prickle—CLASP connection may link how this PCP protein might function at both at the growth cone with Vangl (see discussion above) and at FAs/PCs.

CONVERGENCE EXTENSION DEFECTS LEADING TO OPEN NEURAL TUBE

During vertebrate development, the neural plate (similar to other tissues) undergoes convergent extension to reorganize the cells so that the tissue both narrows and elongates. This involves medial cell intercalation and neighbor exchanges such that two cells positioned laterally to one another but not touching are brought together to create a new junction, and this process proceeds across the entire neural plate (Keller, 2002; Wallingford et al., 2002; Butler and Wallingford, 2017). This intercalation involves apical boundary rearrangement and polarized protrusive behavior at basolateral regions of the cells (Williams et al., 2014), demonstrating a role for lamellipodial protrusions (which as we have seen play key roles in axonal advance and neuronal migration) in convergent extension (Wallingford et al., 2002). Failure to properly execute convergent extension of the neuroepithelium can lead to neural tube defects, and mutations in *VANGL1* and *2, FZD6, PRICKLE1, FUZZY* (a PCP effector), *SCRIB*, and *CELSR1* have been associated with this disorder in humans, ranging from the less severe spina bifida to the complete open neural tube defect craniorachischisis (Kibar et al., 2007, 2009, 2011; Lei et al., 2010; Bosoi et al., 2011; Seo et al., 2011; Allache et al., 2012; Robinson et al., 2012; De Marco et al., 2013). Understanding the mechanisms of neural tube closure has been somewhat hampered by the different processes utilized by *Xenopus*, mice, and humans; however, a recent comprehensive review documents these differences and provides additional insight into the mechanisms that drive neural tube closure (Nikolopoulou et al., 2017). Nonetheless, utilization of PCP and associated components to manipulate cellular processes and growth cone behavior through modulation of the cytoskeleton appears to be a common phenomenon in the nervous system. However, an additional wrinkle that has been employed for neural tube folding results from yet another flavor of actomyosin reorganization and function; namely, that actomyosin contractions along the mediolateral axis at the level of the apical AJs serve to kink the plate to produce neural folds and begin the process of neural tube closure (Nishimura et al., 2012; Ossipova et al., 2015).

CONCLUSIONS

In this chapter, I have summarized evidence showing that PCP function in the nervous system primarily revolves around modulating cytoskeletal dynamics (both the MT and the actin cytoskeletons, acting in concert with one another) at the growth cone of an advancing axon to facilitate the proper pathfinding of that axon, or entire neuron, and intimately coordinating this growth-positive activity with growth-restrictive activities in other regions of the neuron. This process begins with exploratory neurites of a newly born neuron searching for informational cues from the ECM, eventually leading to specification of a single axon. Once an axon is specified, both MTs and actin contribute to protrusive force in the growth cone, controlled by GTPases such as Rac1 and Cdc42 (both effectors of polarity proteins), and FA-like PCs aid in anchoring the cytoskeleton to help provide the necessary force generation. This activity is counterbalanced in minor neurites by the RhoA GTPase (another polarity protein effector) such that PCs are disassembled and stress fibers are activated to retract them. PCs are utilized over and over as axons advance, or neurons migrate, with new ones assembling at the leading edge as older, more rearward ones are disassembled. Polarity pathway genes are involved at all steps of these processes, from reading information in the ECM (or on cells on which the neurons migrate), to actively participating in cytoskeletal and PC dynamics. To fully understand the mechanisms regarding PCP function in neurons, it will be critical to determine the molecular connections between the key players and to precisely determine their subcellular sites of activity.

ACKNOWLEDGMENTS

I would like to express my sincerest thanks to Dr. David Gubb, who continues to be a source of stimulating ideas and unbridled enthusiasm. I sincerely apologize to the polarity community if I have failed to cite any important studies; this was certainly not intentional.

REFERENCES

Akhshi, T.K., Wernike, D., Piekny, A., 2014. Microtubules and actin crosstalk in cell migration and division. Cytoskelet. (Hoboken) 71, 1—23.

Allache, R., De Marco, P., Merello, E., Capra, V., Kibar, Z., 2012. Role of the planar cell polarity gene CELSR1 in neural tube defects and caudal agenesis. Birth Defects Res. A Clin. Mol. Teratol. 94, 176—181.

Andre, P., Wang, Q., Wang, N., Gao, B., Schilit, A., et al., 2012. The Wnt coreceptor Ryk regulates Wnt/planar cell polarity by modulating the degradation of the core planar cell polarity component Vangl2. J. Biol. Chem. 287, 44518—44525.

Arimura, N., Kaibuchi, K., 2007. Neuronal polarity: from extracellular signals to intracellular mechanisms. Nat. Rev. Neurosci. 8, 194—205.

Ayukawa, T., Akiyama, M., Mummery-Widmer, J.L., Stoeger, T., Sasaki, J., et al., 2014. Dachsous-dependent asymmetric localization of spiny-legs determines planar cell polarity orientation in *Drosophila*. Cell Rep. 8, 610—621.

Bard, L., Boscher, C., Lambert, M., Mege, R.M., Choquet, D., et al., 2008. A molecular clutch between the actin flow and N-cadherin adhesions drives growth cone migration. J. Neurosci. 28, 5879—5890.

Bassuk, A.G., Sherr, E.H., 2015. A de novo mutation in PRICKLE1 in fetal agenesis of the corpus callosum and polymicrogyria. J. Neurogenet. 29, 174—177.

Bayly, R., Axelrod, J.D., 2011. Pointing in the right direction: new developments in the field of planar cell polarity. Nat. Rev. Genet. 12, 385—391.

Betschinger, J., Mechtler, K., Knoblich, J.A., 2003. The Par complex directs asymmetric cell division by phosphorylating the cytoskeletal protein Lgl. Nature 422, 326—330.

Boeda, B., Briggs, D.C., Higgins, T., Garvalov, B.K., Fadden, A.J., et al., 2007. Tes, a specific Mena interacting partner, breaks the rules for EVH1 binding. Mol. Cell 28, 1071−1082.

Boeda, B., Knowles, P.P., Briggs, D.C., Murray-Rust, J., Soriano, E., et al., 2011. Molecular recognition of the Tes LIM2-3 domains by the actin-related protein Arp7A. J. Biol. Chem. 286, 11543−11554.

Bosoi, C.M., Capra, V., Allache, R., Trinh, V.Q., De Marco, P., et al., 2011. Identification and characterization of novel rare mutations in the planar cell polarity gene PRICKLE1 in human neural tube defects. Hum. Mutat. 32, 1371−1375.

Boutin, C., Goffinet, A.M., Tissir, F., 2012. Celsr1-3 cadherins in PCP and brain development. Curr. Top. Dev. Biol. 101, 161−183.

Bradke, F., Dotti, C.G., 1999. The role of local actin instability in axon formation. Science 283, 1931−1934.

Breitsprecher, D., Kiesewetter, A.K., Linkner, J., Vinzenz, M., Stradal, T.E., et al., 2011. Molecular mechanism of Ena/VASP-mediated actin-filament elongation. EMBO J. 30, 456−467.

Brieher, W.M., Yap, A.S., 2013. Cadherin junctions and their cytoskeleton(s). Curr. Opin. Cell Biol. 25, 39−46.

Bulgakova, N.A., Knust, E., 2009. The Crumbs complex: from epithelial-cell polarity to retinal degeneration. J. Cell Sci. 122, 2587−2596.

Butler, M.T., Wallingford, J.B., 2017. Planar cell polarity in development and disease. Nat. Rev. Mol. Cell Biol. 18, 375−388.

Carlier, M.F., Pernier, J., Montaville, P., Shekhar, S., Kuhn, S., et al., 2015. Control of polarized assembly of actin filaments in cell motility. Cell Mol. Life Sci. 72, 3051−3067.

Carreira-Barbosa, F., Concha, M.L., Takeuchi, M., Ueno, N., Wilson, S.W., et al., 2003. Prickle 1 regulates cell movements during gastrulation and neuronal migration in zebrafish. Development 130, 4037−4046.

Chen, J., Zhang, M., 2013. The Par3/Par6/aPKC complex and epithelial cell polarity. Exp. Cell Res. 319, 1357−1364.

Cheng, P.L., Lu, H., Shelly, M., Gao, H., Poo, M.M., 2011. Phosphorylation of E3 ligase Smurf1 switches its substrate preference in support of axon development. Neuron 69, 231−243.

Da Silva, J.S., Medina, M., Zuliani, C., Di Nardo, A., Witke, W., et al., 2003. RhoA/ROCK regulation of neuritogenesis via profilin IIa-mediated control of actin stability. J. Cell Biol. 162, 1267−1279.

Davey, C.F., Moens, C.B., 2017. Planar cell polarity in moving cells: think globally, act locally. Development 144, 187−200.

Davey, C.F., Mathewson, A.W., Moens, C.B., 2016. PCP signaling between migrating neurons and their planar-polarized neuroepithelial environment controls filopodial dynamics and directional migration. PLoS Genet. 12, e1005934.

De Marco, P., Merello, E., Consales, A., Piatelli, G., Cama, A., et al., 2013. Genetic analysis of disheveled 2 and disheveled 3 in human neural tube defects. J. Mol. Neurosci. 49, 582−588.

Dent, E.W., Gupton, S.L., Gertler, F.B., 2011. The growth cone cytoskeleton in axon outgrowth and guidance. Cold Spring Harb. Perspect. Biol. 3.

Devenport, D., 2014. The cell biology of planar cell polarity. J. Cell Biol. 207, 171−179.

Devenport, D., 2016. Tissue morphodynamics: translating planar polarity cues into polarized cell behaviors. Semin. Cell Dev. Biol. 55, 99−110.

Djiane, A., Yogev, S., Mlodzik, M., 2005. The apical determinants aPKC and dPatj regulate Frizzled-dependent planar cell polarity in the *Drosophila* eye. Cell 121, 621−631.

Eaton, S., Auvinen, P., Luo, L., Jan, Y.N., Simons, K., 1995. CDC42 and Rac1 control different actin-dependent processes in the *Drosophila* wing disc epithelium. J. Cell Biol. 131, 151−164.

Ehaideb, S.N., Iyengar, A., Ueda, A., Iacobucci, G.J., Cranston, C., et al., 2014. Prickle modulates microtubule polarity and axonal transport to ameliorate seizures in flies. Proc. Natl. Acad. Sci. U.S.A. 111, 11187−11192.

Elsum, I., Yates, L., Humbert, P.O., Richardson, H.E., 2012. The Scribble-Dlg-Lgl polarity module in development and cancer: from flies to man. Essays Biochem. 53, 141−168.

Fanto, M., Weber, U., Strutt, D.I., Mlodzik, M., 2000. Nuclear signaling by Rac and Rho GTPases is required in the establishment of epithelial planar polarity in the *Drosophila* eye. Curr. Biol. 10, 979−988.

Fleming, T., Chien, S.C., Vanderzalm, P.J., Dell, M., Gavin, M.K., et al., 2010. The role of *C. elegans* Ena/VASP homolog UNC-34 in neuronal polarity and motility. Dev. Biol. 344, 94−106.

Flynn, K.C., Hellal, F., Neukirchen, D., Jacob, S., Tahirovic, S., et al., 2012. ADF/cofilin-mediated actin retrograde flow directs neurite formation in the developing brain. Neuron 76, 1091−1107.

Formstone, C.J., Mason, I., 2005. Combinatorial activity of Flamingo proteins directs convergence and extension within the early zebrafish embryo via the planar cell polarity pathway. Dev. Biol. 282, 320−335.

Gao, B., Song, H., Bishop, K., Elliot, G., Garrett, L., et al., 2011. Wnt signaling gradients establish planar cell polarity by inducing Vangl2 phosphorylation through Ror2. Dev. Cell 20, 163−176.

Garcia, M., Leduc, C., Lagardere, M., Argento, A., Sibarita, J.B., et al., 2015. Two-tiered coupling between flowing actin and immobilized N-cadherin/catenin complexes in neuronal growth cones. Proc. Natl. Acad. Sci. U.S.A. 112, 6997−7002.

Garvalov, B.K., Flynn, K.C., Neukirchen, D., Meyn, L., Teusch, N., et al., 2007. Cdc42 regulates cofilin during the establishment of neuronal polarity. J. Neurosci. 27, 13117−13129.

Gerard, A., Mertens, A.E., van der Kammen, R.A., Collard, J.G., 2007. The Par polarity complex regulates Rap1- and chemokine-induced T cell polarization. J. Cell Biol. 176, 863−875.

Gombos, R., Migh, E., Antal, O., Mukherjee, A., Jenny, A., et al., 2015. The formin DAAM functions as molecular effector of the planar cell polarity pathway during axonal development in *Drosophila*. J. Neurosci. 35, 10154−10167.

Goodrich, L.V., Strutt, D., 2011. Principles of planar polarity in animal development. Development 138, 1877−1892.

Goodrich, L.V., 2008. The plane facts of PCP in the CNS. Neuron 60, 9−16.

Gray, R.S., Roszko, I., Solnica-Krezel, L., 2011. Planar cell polarity: coordinating morphogenetic cell behaviors with embryonic polarity. Dev. Cell 21, 120−133.

Griffith, E., Coutts, A.S., Black, D.M., 2005. RNAi knockdown of the focal adhesion protein TES reveals its role in actin stress fibre organisation. Cell Motil. Cytoskelet. 60, 140−152.

Grosse, R., Copeland, J.W., Newsome, T.P., Way, M., Treisman, R., 2003. A role for VASP in RhoA-Diaphanous signalling to actin dynamics and SRF activity. EMBO J. 22, 3050−3061.

Grumolato, L., Liu, G., Mong, P., Mudbhary, R., Biswas, R., et al., 2010. Canonical and noncanonical Wnts use a common mechanism to activate completely unrelated coreceptors. Genes Dev. 24, 2517−2530.

Guo, W., Jiang, H., Gray, V., Dedhar, S., Rao, Y., 2007. Role of the integrin-linked kinase (ILK) in determining neuronal polarity. Dev. Biol. 306, 457−468.

Habas, R., Kato, Y., He, X., 2001. Wnt/Frizzled activation of Rho regulates vertebrate gastrulation and requires a novel Formin homology protein Daam1. Cell 107, 843−854.

Habas, R., Dawid, I.B., He, X., 2003. Coactivation of Rac and Rho by Wnt/Frizzled signaling is required for vertebrate gastrulation. Genes Dev. 17, 295−309.

Harris, T.J., Tepass, U., 2010. Adherens junctions: from molecules to morphogenesis. Nat. Rev. Mol. Cell Biol. 11, 502–514.

Hong, M., Lee, V.M., 1997. Insulin and insulin-like growth factor-1 regulate tau phosphorylation in cultured human neurons. J. Biol. Chem. 272, 19547–19553.

Hur, E.M., Saijilafu, Lee, B.D., Kim, S.J., Xu, W.L., et al., 2011. GSK3 controls axon growth via CLASP-mediated regulation of growth cone microtubules. Genes Dev. 25, 1968–1981.

Hyland, C., Mertz, A.F., Forscher, P., Dufresne, E., 2014. Dynamic peripheral traction forces balance stable neurite tension in regenerating *Aplysia* bag cell neurons. Sci. Rep. 4, 4961.

Insolera, R., Chen, S., Shi, S.H., 2011. Par proteins and neuronal polarity. Dev. Neurobiol. 71, 483–494.

Jessen, J.R., Topczewski, J., Bingham, S., Sepich, D.S., Marlow, F., et al., 2002. Zebrafish trilobite identifies new roles for Strabismus in gastrulation and neuronal movements. Nat. Cell Biol. 4, 610–615.

Jiang, H., Guo, W., Liang, X., Rao, Y., 2005. Both the establishment and the maintenance of neuronal polarity require active mechanisms: critical roles of GSK-3beta and its upstream regulators. Cell 120, 123–135.

Kadrmas, J.L., Beckerle, M.C., 2004. The LIM domain: from the cytoskeleton to the nucleus. Nat. Rev. Mol. Cell Biol. 5, 920–931.

Kallay, L.M., McNickle, A., Brennwald, P.J., Hubbard, A.L., Braiterman, L.T., 2006. Scribble associates with two polarity proteins, Lgl2 and Vangl2, via distinct molecular domains. J. Cell Biochem. 99, 647–664.

Kapitein, L.C., Hoogenraad, C.C., 2011. Which way to go? Cytoskeletal organization and polarized transport in neurons. Mol. Cell. Neurosci. 46, 9–20.

Katoh, K., Kano, Y., Noda, Y., 2011. Rho-associated kinase-dependent contraction of stress fibres and the organization of focal adhesions. J. R. Soc. Interface 8, 305–311.

Keller, R., 2002. Shaping the vertebrate body plan by polarized embryonic cell movements. Science 298, 1950–1954.

Kibar, Z., Torban, E., McDearmid, J.R., Reynolds, A., Berghout, J., et al., 2007. Mutations in VANGL1 associated with neural-tube defects. N. Engl. J. Med. 356, 1432–1437.

Kibar, Z., Bosoi, C.M., Kooistra, M., Salem, S., Finnell, R.H., et al., 2009. Novel mutations in VANGL1 in neural tube defects. Hum. Mutat. 30, E706–E715.

Kibar, Z., Salem, S., Bosoi, C.M., Pauwels, E., De Marco, P., et al., 2011. Contribution of VANGL2 mutations to isolated neural tube defects. Clin. Genet. 80, 76–82.

Kishida, S., Yamamoto, H., Kikuchi, A., 2004. Wnt-3a and Dvl induce neurite retraction by activating Rho-associated kinase. Mol. Cell. Biol. 24, 4487–4501.

Kollins, K.M., Hu, J., Bridgman, P.C., Huang, Y.Q., Gallo, G., 2009. Myosin-II negatively regulates minor process extension and the temporal development of neuronal polarity. Dev. Neurobiol. 69, 279–298.

Kwiatkowski, A.V., Rubinson, D.A., Dent, E.W., Edward van Veen, J., Leslie, J.D., et al., 2007. Ena/VASP Is Required for neuritogenesis in the developing cortex. Neuron 56, 441–455.

Lei, Y.P., Zhang, T., Li, H., Wu, B.L., Jin, L., et al., 2010. VANGL2 mutations in human cranial neural-tube defects. N. Engl. J. Med. 362, 2232–2235.

Li, L., Fothergill, T., Hutchins, B.I., Dent, E.W., Kalil, K., 2014. Wnt5a evokes cortical axon outgrowth and repulsive guidance by tau mediated reorganization of dynamic microtubules. Dev. Neurobiol. 74, 797–817.

Lim, B.C., Matsumoto, S., Yamamoto, H., Mizuno, H., Kikuta, J., et al., 2016. Prickle1 promotes focal adhesion disassembly in cooperation with the CLASP-LL5beta complex in migrating cells. J. Cell Sci. 129, 3115–3129.

Lu, W., Fox, P., Lakonishok, M., Davidson, M.W., Gelfand, V.I., 2013. Initial neurite outgrowth in *Drosophila* neurons is driven by kinesin-powered microtubule sliding. Curr. Biol. 23, 1018–1023.

Lu, Q., Schafer, D.A., Adler, P.N., 2015. The *Drosophila* planar polarity gene multiple wing hairs directly regulates the actin cytoskeleton. Development 142, 2478–2486.

Lyuksyutova, A.I., Lu, C.C., Milanesio, N., King, L.A., Guo, N., et al., 2003. Anterior-posterior guidance of commissural axons by Wnt-Frizzled signaling. Science 302, 1984–1988.

Mapp, O.M., Wanner, S.J., Rohrschneider, M.R., Prince, V.E., 2010. Prickle1b mediates interpretation of migratory cues during zebrafish facial branchiomotor neuron migration. Dev. Dyn. 239, 1596–1608.

Marlow, F., Topczewski, J., Sepich, D., Solnica-Krezel, L., 2002. Zebrafish Rho kinase 2 acts downstream of Wnt11 to mediate cell polarity and effective convergence and extension movements. Curr. Biol. 12, 876–884.

Matusek, T., Djiane, A., Jankovics, F., Brunner, D., Mlodzik, M., et al., 2006. The *Drosophila* formin DAAM regulates the tracheal cuticle pattern through organizing the actin cytoskeleton. Development 133, 957–966.

Mitchison, T., Kirschner, M., 1988. Cytoskeletal dynamics and nerve growth. Neuron 1, 761–772.

Montcouquiol, M., Rachel, R.A., Lanford, P.J., Copeland, N.G., Jenkins, N.A., et al., 2003. Identification of Vangl2 and Scrb1 as planar polarity genes in mammals. Nature 423, 173–177.

Moon, H.M., Wynshaw-Boris, A., 2013. Cytoskeleton in action: lissencephaly, a neuronal migration disorder. Wiley Interdiscip. Rev. Dev. Biol. 2, 229–245.

Mrkusich, E.M., Flanagan, D.J., Whitington, P.M., 2011. The core planar cell polarity gene prickle interacts with flamingo to promote sensory axon advance in the *Drosophila* embryo. Dev. Biol. 358, 224–230.

Munoz-Descalzo, S., Gomez-Cabrero, A., Mlodzik, M., Paricio, N., 2007. Analysis of the role of the Rac/Cdc42 GTPases during planar cell polarity generation in *Drosophila*. Int. J. Dev. Biol. 51, 379–387.

Myers, J.P., Santiago-Medina, M., Gomez, T.M., 2011. Regulation of axonal outgrowth and pathfinding by integrin-ECM interactions. Dev. Neurobiol. 71, 901–923.

Nagaoka, T., Inutsuka, A., Begum, K., Bin hafiz, K., Kishi, M., 2014. Vangl2 regulates E-cadherin in epithelial cells. Sci. Rep. 4, 6940.

Ng, J., 2012. Wnt/PCP proteins regulate stereotyped axon branch extension in *Drosophila*. Development 139, 165–177.

Nikolopoulou, E., Galea, G.L., Rolo, A., Greene, N.D., Copp, A.J., 2017. Neural tube closure: cellular, molecular and biomechanical mechanisms. Development 144, 552–566.

Nishimura, T., Yamaguchi, T., Kato, K., Yoshizawa, M., Nabeshima, Y., et al., 2005. PAR-6-PAR-3 mediates cdc42-induced rac activation through the rac GEFs STEF/Tiam1. Nat. Cell Biol. 7, 270–277.

Nishimura, T., Honda, H., Takeichi, M., 2012. Planar cell polarity links axes of spatial dynamics in neural-tube closure. Cell 149, 1084–1097.

Nusse, R., 2012. Wnt signaling. Cold Spring Harb. Perspect. Biol. 4.

Olofsson, J., Sharp, K.A., Matis, M., Cho, B., Axelrod, J.D., 2014. Prickle/spiny-legs isoforms control the polarity of the apical microtubule network in planar cell polarity. Development 141, 2866–2874.

Onishi, K., Shafer, B., Lo, C., Tissir, F., Goffinet, A.M., et al., 2013. Antagonistic functions of Dishevelleds regulate Frizzled3 endocytosis via filopodia tips in Wnt-mediated growth cone guidance. J. Neurosci. 33, 19071–19085.

Ossipova, O., Kim, K., Sokol, S.Y., 2015. Planar polarization of Vangl2 in the vertebrate neural plate is controlled by Wnt and Myosin II signaling. Biol. Open 4, 722−730.

Padmanabhan, A., Rao, M.V., Wu, Y., Zaidel-Bar, R., 2015. Jack of all trades: functional modularity in the adherens junction. Curr. Opin. Cell Biol. 36, 32−40.

Pertz, O., 2010. Spatio-temporal Rho GTPase signaling − where are we now? J. Cell Sci. 123, 1841−1850.

Pollard, T.D., Borisy, G.G., 2003. Cellular motility driven by assembly and disassembly of actin filaments. Cell 112, 453−465.

Rivas, R.J., Hatten, M.E., 1995. Motility and cytoskeletal organization of migrating cerebellar granule neurons. J. Neurosci. 15, 981−989.

Robinson, A., Escuin, S., Doudney, K., Vekemans, M., Stevenson, R.E., et al., 2012. Mutations in the planar cell polarity genes CELSR1 and SCRIB are associated with the severe neural tube defect craniorachischisis. Hum. Mutat. 33, 440−447.

Roossien, D.H., Lamoureux, P., Van Vactor, D., Miller, K.E., 2013. *Drosophila* growth cones advance by forward translocation of the neuronal cytoskeletal meshwork in vivo. PLoS One 8, e80136.

Sato, M., Umetsu, D., Murakami, S., Yasugi, T., Tabata, T., 2006. DWnt4 regulates the dorsoventral specificity of retinal projections in the *Drosophila melanogaster* visual system. Nat. Neurosci. 9, 67−75.

Schaefer, A.W., Schoonderwoert, V.T., Ji, L., Mederios, N., Danuser, G., et al., 2008. Coordination of actin filament and microtubule dynamics during neurite outgrowth. Dev. Cell 15, 146−162.

Schelski, M., Bradke, F., 2017 (in press). Neuronal polarization: from spatiotemporal signaling to cytoskeletal dynamics. Mol. Cell Neurosci.

Schwamborn, J.C., Puschel, A.W., 2004. The sequential activity of the GTPases Rap1B and Cdc42 determines neuronal polarity. Nat. Neurosci. 7, 923−929.

Seetapun, D., Odde, D.J., 2010. Cell-length-dependent microtubule accumulation during polarization. Curr. Biol. 20, 979−988.

Seo, J.H., Zilber, Y., Babayeva, S., Liu, J., Kyriakopoulos, P., et al., 2011. Mutations in the planar cell polarity gene, Fuzzy, are associated with neural tube defects in humans. Hum. Mol. Genet. 20, 4324−4333.

Shafer, B., Onishi, K., Lo, C., Colakoglu, G., Zou, Y., 2011. Vangl2 promotes Wnt/planar cell polarity-like signaling by antagonizing Dvl1-mediated feedback inhibition in growth cone guidance. Dev. Cell 20, 177−191.

Shakir, M.A., Gill, J.S., Lundquist, E.A., 2006. Interactions of UNC-34 enabled with Rac GTPases and the NIK kinase MIG-15 in *Caenorhabditis elegans* axon pathfinding and neuronal migration. Genetics 172, 893−913.

Shi, S.H., Cheng, T., Jan, L.Y., Jan, Y.N., 2004. APC and GSK-3beta are involved in mPar3 targeting to the nascent axon and establishment of neuronal polarity. Curr. Biol. 14, 2025−2032.

Shimizu, K., Sato, M., Tabata, T., 2011. The Wnt5/planar cell polarity pathway regulates axonal development of the *Drosophila* mushroom body neuron. J. Neurosci. 31, 4944−4954.

Short, C.A., Suarez-Zayas, E.A., Gomez, T.M., 2016. Cell adhesion and invasion mechanisms that guide developing axons. Curr. Opin. Neurobiol. 39, 77−85.

Simons, M., Mlodzik, M., 2008. Planar cell polarity signaling: from fly development to human disease. Annu. Rev. Genet. 42, 517−540.

Smith, M.A., Hoffman, L.M., Beckerle, M.C., 2014. LIM proteins in actin cytoskeleton mechanoresponse. Trends Cell Biol. 24, 575−583.

Sokol, S.Y., 2015. Spatial and temporal aspects of Wnt signaling and planar cell polarity during vertebrate embryonic development. Semin. Cell Dev. Biol. 42, 78−85.

Spiering, D., Hodgson, L., 2011. Dynamics of the Rho-family small GTPases in actin regulation and motility. Cell Adhes. Migr. 5, 170–180.

Stehbens, S.J., Paszek, M., Pemble, H., Ettinger, A., Gierke, S., et al., 2014. CLASPs link focal-adhesion-associated microtubule capture to localized exocytosis and adhesion site turnover. Nat. Cell Biol. 16, 561–573.

Strutt, D.I., Weber, U., Mlodzik, M., 1997. The role of RhoA in tissue polarity and Frizzled signalling. Nature 387, 292–295.

Su, W.H., Mruk, D.D., Wong, E.W., Lui, W.Y., Cheng, C.Y., 2012. Polarity protein complex Scribble/Lgl/Dlg and epithelial cell barriers. Adv. Exp. Med. Biol. 763, 149–170.

Suter, D.M., Errante, L.D., Belotserkovsky, V., Forscher, P., 1998. The Ig superfamily cell adhesion molecule, apCAM, mediates growth cone steering by substrate-cytoskeletal coupling. J. Cell Biol. 141, 227–240.

Tahinci, E., Symes, K., 2003. Distinct functions of Rho and Rac are required for convergent extension during *Xenopus* gastrulation. Dev. Biol. 259, 318–335.

Tahirovic, S., Hellal, F., Neukirchen, D., Hindges, R., Garvalov, B.K., et al., 2010. Rac1 regulates neuronal polarization through the WAVE complex. J. Neurosci. 30, 6930–6943.

Takano, T., Xu, C., Funahashi, Y., Namba, T., Kaibuchi, K., 2015. Neuronal polarization. Development 142, 2088–2093.

Takenawa, T., Suetsugu, S., 2007. The WASP-WAVE protein network: connecting the membrane to the cytoskeleton. Nat. Rev. Mol. Cell Biol. 8, 37–48.

Thrasivoulou, C., Millar, M., Ahmed, A., 2013. Activation of intracellular calcium by multiple Wnt ligands and translocation of beta-catenin into the nucleus: a convergent model of Wnt/Ca2+ and Wnt/beta-catenin pathways. J. Biol. Chem. 288, 35651–35659.

Tissir, F., Goffinet, A.M., 2013. Shaping the nervous system: role of the core planar cell polarity genes. Nat. Rev. Neurosci. 14, 525–535.

Toriyama, M., Shimada, T., Kim, K.B., Mitsuba, M., Nomura, E., et al., 2006. Shootin1: a protein involved in the organization of an asymmetric signal for neuronal polarization. J. Cell Biol. 175, 147–157.

Trivedi, N., Marsh, P., Goold, R.G., Wood-Kaczmar, A., Gordon-Weeks, P.R., 2005. Glycogen synthase kinase-3beta phosphorylation of MAP1B at Ser1260 and Thr1265 is spatially restricted to growing axons. J. Cell Sci. 118, 993–1005.

Tsai, J.W., Bremner, K.H., Vallee, R.B., 2007. Dual subcellular roles for LIS1 and dynein in radial neuronal migration in live brain tissue. Nat. Neurosci. 10, 970–979.

Tsuji, T., Ohta, Y., Kanno, Y., Hirose, K., Ohashi, K., et al., 2010. Involvement of p114-RhoGEF and Lfc in Wnt-3a- and dishevelled-induced RhoA activation and neurite retraction in N1E-115 mouse neuroblastoma cells. Mol. Biol. Cell 21, 3590–3600.

Wada, H., Okamoto, H., 2009. Roles of planar cell polarity pathway genes for neural migration and differentiation. Dev. Growth Differ. 51, 233–240.

Wada, H., Iwasaki, M., Sato, T., Masai, I., Nishiwaki, Y., et al., 2005. Dual roles of zygotic and maternal Scribble1 in neural migration and convergent extension movements in zebrafish embryos. Development 132, 2273–2285.

Wada, H., Tanaka, H., Nakayama, S., Iwasaki, M., Okamoto, H., 2006. Frizzled3a and Celsr2 function in the neuroepithelium to regulate migration of facial motor neurons in the developing zebrafish hindbrain. Development 133, 4749–4759.

Wallingford, J.B., Fraser, S.E., Harland, R.M., 2002. Convergent extension: the molecular control of polarized cell movement during embryonic development. Dev. Cell 2, 695–706.

Wallingford, J.B., 2012. Planar cell polarity and the developmental control of cell behavior in vertebrate embryos. Annu. Rev. Cell Dev. Biol. 28, 627–653.

Walsh, G.S., Grant, P.K., Morgan, J.A., Moens, C.B., 2011. Planar polarity pathway and Nance-Horan syndrome-like 1b have essential cell-autonomous functions in neuronal migration. Development 138, 3033–3042.

Wansleeben, C., Meijlink, F., 2011. The planar cell polarity pathway in vertebrate development. Dev. Dyn. 240, 616–626.

Warrington, S.J., Strutt, H., Strutt, D., 2013. The Frizzled-dependent planar polarity pathway locally promotes E-cadherin turnover via recruitment of RhoGEF2. Development 140, 1045–1054.

Watabe-Uchida, M., John, K.A., Janas, J.A., Newey, S.E., Van Aelst, L., 2006. The Rac activator DOCK7 regulates neuronal polarity through local phosphorylation of stathmin/Op18. Neuron 51, 727–739.

Whitford, K.L., Dijkhuizen, P., Polleux, F., Ghosh, A., 2002. Molecular control of cortical dendrite development. Annu. Rev. Neurosci. 25, 127–149.

Williams, B.B., Cantrell, V.A., Mundell, N.A., Bennett, A.C., Quick, R.E., et al., 2012. VANGL2 regulates membrane trafficking of MMP14 to control cell polarity and migration. J. Cell Sci. 125, 2141–2147.

Williams, M., Yen, W., Lu, X., Sutherland, A., 2014. Distinct apical and basolateral mechanisms drive planar cell polarity-dependent convergent extension of the mouse neural plate. Dev. Cell 29, 34–46.

Winograd-Katz, S.E., Fassler, R., Geiger, B., Legate, K.R., 2014. The integrin adhesome: from genes and proteins to human disease. Nat. Rev. Mol. Cell Biol. 15, 273–288.

Winter, C.G., Wang, B., Ballew, A., Royou, A., Karess, R., et al., 2001. *Drosophila* Rho-associated kinase (Drok) links Frizzled-mediated planar cell polarity signaling to the actin cytoskeleton. Cell 105, 81–91.

Witte, H., Bradke, F., 2008. The role of the cytoskeleton during neuronal polarization. Curr. Opin. Neurobiol. 18, 479–487.

Xie, Z., Sanada, K., Samuels, B.A., Shih, H., Tsai, L.H., 2003. Serine 732 phosphorylation of FAK by Cdk5 is important for microtubule organization, nuclear movement, and neuronal migration. Cell 114, 469–482.

Yang, Y., Mlodzik, M., 2015. Wnt-Frizzled/planar cell polarity signaling: cellular orientation by facing the wind (Wnt). Annu. Rev. Cell Dev. Biol. 31, 623–646.

Yau, K.W., Schatzle, P., Tortosa, E., Pages, S., Holtmaat, A., et al., 2016. Dendrites in vitro and in vivo contain microtubules of opposite polarity and axon formation correlates with uniform plus-end-out microtubule orientation. J. Neurosci. 36, 1071–1085.

Yoshimura, T., Arimura, N., Kawano, Y., Kawabata, S., Wang, S., et al., 2006. Ras regulates neuronal polarity via the PI3-kinase/Akt/GSK-3beta/CRMP-2 pathway. Biochem. Biophys. Res. Commun. 340, 62–68.

Yu, W., Baas, P.W., 1994. Changes in microtubule number and length during axon differentiation. J. Neurosci. 14, 2818–2829.

Zaidel-Bar, R., 2013. Cadherin adhesome at a glance. J. Cell Sci. 126, 373–378.

Zhang, L., Luga, V., Armitage, S.K., Musiol, M., Won, A., et al., 2016. A lateral signalling pathway coordinates shape volatility during cell migration. Nat. Commun. 7, 11714.

Zou, Y., 2012. Does planar cell polarity signaling steer growth cones? Curr. Top. Dev. Biol. 101, 141–160.

Zumbrunn, J., Kinoshita, K., Hyman, A.A., Nathke, I.S., 2001. Binding of the adenomatous polyposis coli protein to microtubules increases microtubule stability and is regulated by GSK3 beta phosphorylation. Curr. Biol. 11, 44–49.

Chapter 7

Planar Cell Polarity in Ciliated Epithelia

Peter Walentek[1], Camille Boutin[2], Laurent Kodjabachian[2]
[1]*University of California, Berkeley, CA, United States;* [2]*Aix-Marseille University, Marseille, France*

INTRODUCTION

Polarity can be visualized at all organizational scales of living matter, from molecules, to organelles, to individual cells, to groups of cells or tissues, up to the entire organism (reviewed in e.g., Satir, 2016; Vladar et al., 2009; Wang and Nathans, 2007). Correct coordination between those successive levels of organization is necessary for the emergence of physiological functions. In this chapter, we shall describe how such coordination is ensured in ciliated epithelia, as they offer excellent paradigms to study polarity across the entire spectrum of scales.

In all animals, tissues and organs are compartmentalized through the presence of epithelia at their surface, which form selective barriers with the external or internal environment (St Johnston and Ahringer, 2010). Epithelia are endowed with basic roles such as absorption, filtration, and secretion, which mainly depend on apicobasal cell polarity. Apicobasal polarity allows the intercellular formation of tight junctions and controlled sorting of biological materials from or toward the apical and basal cell membrane (Van Itallie and Anderson, 2014). Epithelia can also be specialized to ensure additional functions, such as the vectorial transport of particles or cells along their surface via coordinated beating of motile cilia (Marshall and Kintner, 2008). Here, polarity information is distributed along the plane of the tissue and organizes the apical domains of ciliated epithelial cells. In this chapter, we focus on planar cell polarity (PCP) of ciliated epithelia, although coordination between apicobasal and planar polarity systems will certainly prove relevant in the future.

When considering ciliated epithelia, we may distinguish those that harbor cells bearing a single motile cilium from those containing multiciliated cells (MCCs) (Ibanez-Tallon et al., 2003). In both types of ciliated epithelia, planar

Cell Polarity in Development and Disease. https://doi.org/10.1016/B978-0-12-802438-6.00007-3
177

polarity is essential for the production of directional fluid flow, through the control of cilia polarization and beating orientation (Wallingford, 2010). For instance, it is known that PCP pathway components are essential for the orientation of single cilia in cells of the vertebrate embryonic left—right organizer. Vast literature including excellent reviews exist on this subject (e.g., Blum et al., 2014; Brown and Wolpert, 1990; Marshall and Nonaka, 2006; Schweickert et al., 2011), thus, we shall focus in this chapter on ciliated epithelia harboring MCCs (Fig. 7.1).

MCC-containing ciliated epithelia are present throughout metazoan evolution and serve functions ranging from locomotion of marine larvae and flatworms to brain homeostasis, mucociliary clearance of pathogens, and transportation of oocytes in vertebrates (Ganesan et al., 2013; Ibanez-Tallon et al., 2003; Meunier and Azimzadeh, 2016; Rompolas et al., 2009). As classical invertebrate model organisms lack MCCs, much of our current knowledge comes from studies performed with frogs, mice, human tissue culture, and, to a lesser extent, fish and chicken. In this chapter, we will cover three paradigms that have all contributed extensively to our understanding of how polarity pathways organize multiciliated epithelia: the *Xenopus* embryonic epidermis, the mouse tracheobronchial epithelium, and the mouse ependyma (Fig. 7.1). These three models share common features but also display differences that make their comparison particularly instructive.

The *Xenopus* embryonic epidermis comprises a mucociliary epithelium, which is thought to both help fighting potential pathogens and to allow oxygenation through the skin via cilia-driven fluid flow at the surface of the embryo (Fig. 7.1A) (Hayes et al., 2007; Werner and Mitchell, 2012). This integrated functional unit is composed of: (1) a majority of goblet-like mucus-secreting cells, which produce and release mucin-like substances (e.g., Otogelin), as well as peptides with antimicrobial properties (e.g., Intelectin); (2) MCCs, which account for about 20% of all cells, are regularly spaced over the entire surface of the embryo, each containing nearly 200 motile cilia (Fig. 7.1B—C); (3) ion-secreting cells (ionocytes), which control pH homeostasis; and (4) small secretory cells (SSCs), which secrete antimicrobial substances and release serotonin that controls ciliary beating frequency in MCCs (Dubaissi and Papalopulu, 2011; Dubaissi et al., 2014; Hayes et al., 2007; Quigley et al., 2011; Walentek et al., 2014). This epithelium overlies a non-cohesive cell layer made of stem-like precursor cells that will be mobilized at metamorphosis to build the definitive pluristratified skin of the adult. Through the ease of functional manipulations and analysis in *Xenopus*, the embryonic mucociliary epidermis model has delivered the largest body of information in the field of MCC formation (Brooks and Wallingford, 2014; Werner and Mitchell, 2012).

The mammalian tracheobronchial epithelium represents another relevant model to study various aspects of mucociliary biology (Fig. 7.1D—E). Here, the predominant in vivo model is the mouse tracheal epithelium that can be

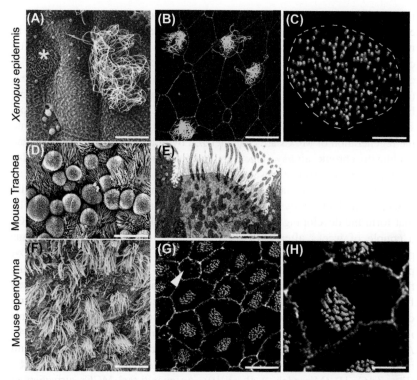

FIGURE 7.1 *Xenopus* **and mouse multiciliated epithelia.** (A) Scanning electron microscopy (SEM) micrograph depicting the various cell types of the *Xenopus* embryonic mucociliary epidermis. Asterisks: multiciliated cell (MCC) (green), Goblet-like cell (red), Small Secretory Cell (brown), Ionocyte (Ion secreting cells, yellow). Scale bar: 5 μm. (B) Immunofluorescent micrograph of the *Xenopus* embryonic epidermis stained for the cell membrane marker ZO1 (white) and the cilia marker acetyl-α-tubulin (green). Scale bar: 10 μm. (C) Fluorescent micrograph of a *Xenopus* MCC stained for γ-tubulin (red) and Centrin4 (green) revealing basal body (BB) organization and polarity. The dashed line indicates the cell membrane. BBs are dispersed evenly over the entire apical surface and their orientation is uniform. Scale bar: 2.5 μm (D) SEM micrograph of mouse trachea. Asterisks: MCC (green); Club secretory cell (purple). Scale bar: 10 μm. (E) Transmission Electron Microscopy micrograph of a transverse section of a mouse tracheal MCC, revealing BBs docked at the membrane and the extracellular ciliary axonemes. Scale bar: 5 μm (F) SEM micrograph of mouse ependymal wall, which is mainly composed of MCCs (green asterisk). Scale bar: 10 μm. (G) Fluorescent micrograph of the mouse ependymal wall stained with the membrane marker ZO1 (white), centriolar marker FGFR1 oncogene partner (green), and basal foot marker γ-tubulin (red). BBs are clustered in apical patches, which are located off-center. Scale bar: 15 μm. Note the presence of rare neural stem cells in the epithelium (*white arrowhead*). (H) Higher magnification of an MCC from (G). Within a patch, BBs are organized in parallel rows with a similar orientation. Scale bar: 5 μm.

also studied in tissue culture (Rock et al., 2010; Vladar and Brody, 2013). Although adult primary human airway epithelial cell cultures may represent a more relevant model from a biomedical point of view, they turn out to be much more resistant to functional manipulations, and prevent comparison to the in vivo situation (Fulcher et al., 2005). Similar to the *Xenopus* mucociliary epithelium, the mouse pseudostratified tracheal epithelium is comprised of mucus-secreting goblet cells, various additional secretory cells (e.g., Club cells), and MCCs at high density (unlike in *Xenopus*) containing hundreds of cilia, and basal stem cells that have the capacity to (re-) generate all cell types (Fig. 7.1D) (Hogan et al., 2014). Similar to *Xenopus*, secretory cells provide a barrier against pathogens and pollutants, and MCCs provide the force to expel soiled mucus from the airways. Thus, mutations affecting the development and the regeneration of MCCs are characterized by reduced mucociliary clearance leading to chronic airway infections and inflammation (Boon et al., 2014; Wallmeier et al., 2014; Zariwala et al., 2011).

Finally, the ependymal walls of the neonate mouse brain provide another attractive model to study multiciliated epithelia. Indeed, the majority of cells that form the developing epithelium will eventually differentiate into MCCs a couple of weeks after birth (Spassky, 2005) (Fig. 7.1F−H). It is thus possible to manipulate gene function in MCCs through electroporation of plasmids in their progenitors in newly born pups (Boutin et al., 2008). Like the *Xenopus* embryonic epidermis, this model is also particularly well suited for high-resolution and live-cell imaging, and the progressive differentiation of MCCs along the anterior-posterior (AP) axis of the brain allows visualization of all steps of the process in one given animal (Al Jord et al., 2014). Another specific feature of ependymal MCCs is the fact that cilia are found in a patch of about 40−50 units, unlike in *Xenopus* or tracheal MCCs (Fig. 7.1G−H) (Mirzadeh et al., 2010). As ependymal MCCs also display secretory activity, they may represent an interesting case of integrated mucociliary function. Ependymal MCCs generate cerebrospinal fluid circulation in the central nervous system and may also play a significant role in neural stem cell homeostasis (Lim and Alvarez-Buylla, 2014; Sawamoto et al., 2006).

In all models, MCC specification and ciliogenesis appear to occur through a conserved sequence of events regulated by common molecular pathways. After cell cycle exit, the nuclear factors GemC1 and Multicilin activate an MCC-specific transcriptional program, which allows massive centriole amplification, ciliogenesis, and emergence of motility (Brooks and Wallingford, 2014; Choksi et al., 2014; Walentek and Quigley, 2017). Centrioles mature into basal bodies (BBs) through the addition of appendages that allow them to anchor at the apical cell membrane. Following docking, BBs serve as bases for cilia formation and as knobs to rotate cilia in the right direction (Garcia and Reiter, 2016; Marshall, 2008; Zhang and Mitchell, 2016).

In this chapter, we will review data that illustrate why ciliated epithelia represent unique models to understand how polarity is transmitted across scales from the organism level down to individual organelles. We will present that PCP proteins (introduced in the Preface and in Chapters 4 and 6) are involved at all levels of organization and in all models. We will also point out important missing links and open questions that might become the focus of future studies to improve our current understanding.

INITIATION AND COORDINATION OF TISSUE-LEVEL PLANAR POLARITY IN CILIATED EPITHELIA

Generally speaking, establishment and maintenance of tissue level planar polarity requires integration of top-down and bottom-up informational cues (Devenport, 2014; Goodrich and Strutt, 2011; Vladar et al., 2009; Wang and Nathans, 2007). Specifically, every organ and tissue needs to coordinate its polarity within the global directional and morphological framework of the animal to fulfill its particular function. Each cell displays polarity at the molecular level, which is aligned with the tissue-level polarity. Thus, tissue-level polarity could, in principle, be either initiated at the whole embryo level and be passed on to the tissue level, or arise first at the level of one or more individual cells, and be subsequently propagated throughout the tissue. Alternatively, tissue-level polarity could be generated though a complex interplay and bidirectional feedback between those layers of polarity. The latter possibility is certainly attractive, since it would allow the highest degree of flexibility to accommodate for the vast variety of tissue shapes and functions found in the animal kingdom. In any case, the central question remains: What is the very first step of PCP establishment, or what cue tips the balance initially? This fascinating question is subject to debate ever since the first discovery of planar polarity mechanisms in the fruit fly. Over the past years, several seminal studies have started to shed light on these issues, which will be discussed in this section in the context of ciliated epithelia.

Xenopus Embryonic Epidermis

Setting the Stage for Planar Cell Polarity in the Mucociliary Embryonic Epidermis of Xenopus

At the cellular level, PCP employs many molecular components linked to the Wnt signaling pathway, widely studied in the context of gene expression regulation, such as transmembrane Frizzled (Fzd) receptors and Dishevelled-like (Dvl) cytoplasmic mediators (Yang and Mlodzik, 2015). The canonical branch of Wnt signaling (also called Wnt/β-catenin signaling) is activated by binding of Wnt ligands to Fzd and by subsequent Fzd-dependent phosphorylation of Dvl (Gao and Chen, 2010). In contrast to the situation in the fruit fly, where Wnt ligands could not be linked to core PCP signaling by authoritative experiments until recently, it was clear early on that in vertebrate PCP signaling a subset of Wnt ligands, especially Wnt11b and Wnt5a, plays central and instructive roles in planar polarization of the embryo (Andre et al., 2015; Cha et al., 2008; Hardy et al., 2008; Tada and Smith, 2000; Tada et al., 2002; Wu et al., 2013). Therefore, PCP signaling in vertebrates is most commonly referred to as Wnt/PCP signaling to highlight its intimate connection to the Wnt pathway (Yang and Mlodzik, 2015).

In vertebrate development, global establishment of planar polarity is tightly linked to the establishment of the primary embryonic axes, i.e., the AP and dorsal-ventral axes. Interestingly, this process is largely driven by canonical Wnt signaling in an evolutionarily conserved manner (Hikasa and Sokol, 2013). This principle is best illustrated using the early development of amphibian embryos as an example, where Wnt ligands and intracellular components of the Wnt/β-catenin pathway are translocated within the fertilized egg in response to sperm entry (Fig. 7.2-st.0). In a process called cortical rotation, those components accumulate in a defined area, which ultimately induces the formation of the dorsal Spemann−Mangold organizer (Houston, 2012; Spemann and Mangold, 1924). In subsequent steps, the organizer orchestrates regionalized gene expression and morphogenetic movements through the production of secreted signaling molecules and soluble antagonists acting throughout the embryo (Fig. 7.2-st.8−10.5) (Heasman, 2006). Factors leading to organizer formation in *Xenopus* include the maternal ligand Wnt11b, which was shown to regulate Wnt/PCP signaling, in addition to canonical and other noncanonical Wnt pathway branches (Cha et al., 2008). Furthermore, zygotic Wnt11b (as well as Wnt5a) is prominently expressed in a posterior domain during *Xenopus* gastrulation and neurulation (Fig. 7.2-st.10.5−16), and it was the first Wnt ligand demonstrated to functionally regulate PCP-dependent morphogenetic movements (Smith et al., 2000). Later on, Wnt11b was also shown to be required for the establishment of left−right asymmetry by regulation of planar alignment of cilia and morphogenesis of the left−right organizer (Walentek et al., 2013). These findings have prompted speculation that Wnt11b and Wnt5a could form a gradient along the AP axis (with high ligand levels posteriorly), which could set global polarity cues throughout the embryo and initiate PCP in the mucociliary epidermis. Such a scenario would fit the general concept of morphogen gradients in development and provide vertebrates with a gradient-dependent mechanism for orienting PCP similar to the Fat−Dachsous (Ft−Ds) system of cadherins in the fly (Devenport, 2014).

The Ft−Ds pathway is characterized by the formation of two opposing gradients of dachsous and the kinase Four-jointed (Fj) at the tissue level, which influence core PCP components at the cellular level. It is noteworthy that the Ft−Ds pathway is thought to act, at least in part, in a redundant manner to the core PCP pathway, and the relative alignment of Ft−Ds to core PCP components is not strict. This is exemplified by the finding that Ds−Fj gradients are oriented in opposite directions relative to core PCP components in the developing fly wing and other parts of the body (e.g., the eyes) (Devenport, 2014). Furthermore, recent studies in flies have meanwhile demonstrated the instructive role of Wnt ligands for core PCP alignment (Wu et al., 2013), and studies in mice and zebrafish indicate roles for the Ft−Ds pathway in

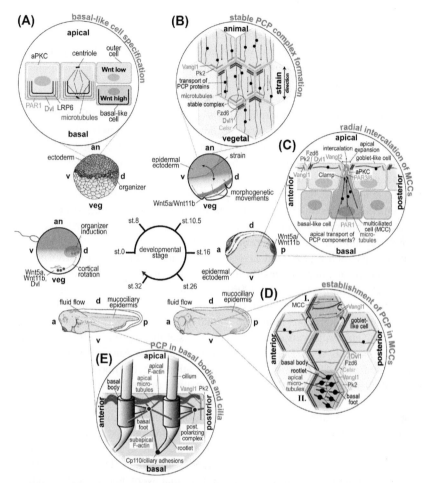

FIGURE 7.2   Planar cell polarity (PCP)–dependent processes in the mucociliary embryonic epidermis of *Xenopus*. Subsequent developmental stages are depicted in the center and specific processes during these stages are indicated. Panels (A)–(E) illustrate specific processes during epidermal development, which rely on Wnt/PCP signaling. (A) During blastula/gastrula stages, epidermal cells divide asymmetrically to generate outer cells and basal-like cells of the epidermis. (B) Morphogenetic movements during gastrulation generate mechanical strain on epidermal cells, which initiates polarized microtubule formation. Polarized transport of PCP components along these microtubules leads to the initiation of stable PCP complexes at lateral membranes. (C) Radial intercalation of deep layer cells (such as multiciliated cell [MCCs]) is dependent on PCP signaling as well as input from the PAR signaling pathway. (D) The establishment of PCP in MCCs requires multiple processes. (D-I) First, formation of asymmetric PCP complexes at cell membranes of MCCs is driven by noncell autonomous input from adjacent goblet-like cells through heterophilic interactions of core PCP components across membranes. Additionally, cell autonomous mechanisms (e.g., Pk2-dependent stabilization of Vangl1 or inhibition of Fzd/Dvl) further refine and strengthen PCP information in the cell. (D-II) Second, cellular PCP information is transmitted to individual basal bodies via microtubules, which interact with basal feet of basal bodies to induce uniform orientation of cilia. (E) The initial PCP-dependent alignment of MCC basal bodies is further refined through flow- and subapical actin-dependent mechanisms that involve PCP components, and which facilitate metachronal synchronous cilia beating.

vertebrate planar polarity-dependent developmental events (Le Pabic et al., 2014; Mao et al., 2011). Therefore, it is expected that future experiments will elucidate the precise roles of Wnt ligands in flies and Ft—Ds signaling in vertebrates, which may allow us to develop a unified model of planar polarity signaling in animals.

Only very recently has the connection between primary axis formation, Wnt ligand expression, morphogenesis, and establishment of PCP in the mucociliary epidermis during *Xenopus* development begun to be elucidated. In a set of elegant experiments, Chien et al. (2015) have shown that the morphogenetic movements, which drive involution of the mesendoderm during gastrulation, provide the necessary cue to bias asymmetries of PCP components along the AP axis (Fig. 7.2-st.10.5). As a result of these cell movements, the ectoderm experiences mechanical strain, i.e., stretching of epidermal precursors. This causes alignment of apical noncentrosomal microtubules along the axis of strain in ectodermal cells (Fig. 7.2B). Core PCP components such as Fzd and Dvl are known to be directionally transported along subapical microtubules during establishment of PCP in the fly (Matis et al., 2014; Shimada et al., 2006), and this seems to be the case in the mucociliary epidermis as well (Chien et al., 2015). Mechanical strain at orthogonal cell junctions along the animal—vegetal (An/Vg) axis of the early embryonic epidermis causes the formation of stabilized protein clusters of core PCP components (Fig. 7.2B). In a simplified view, the An/Vg polarity is then transposed into AP polarity during gastrulation due to morphogenetic reorganization of the embryo. Thus, the formation of stable core PCP clusters along the An/Vg axis provides the initial cue for PCP along the AP axis. Importantly, the authors demonstrated that mechanical strain is not sufficient to establish and maintain PCP, and that polarity is lost when core PCP components are knocked down (Chien et al., 2015). Conversely, without mechanical strain and subapical microtubule alignment, PCP can be established eventually (e.g., in explanted tissues that were not subject to forces generated by gastrulation movements), but the alignment is delayed and poorer (Chien et al., 2015).

Taken together, after induction of the organizer by canonical Wnt/β-catenin signaling, spatially coordinated gene expression of noncanonical Wnt ligands is established, which in turn regulates gastrulation movements. Morphogenetic movements exert mechanical strain on epidermal cells, which leads to alignment of subapical microtubules along the AP axis. Core PCP components are transported in a directional fashion along these microtubules to establish the initial stable fraction of core PCP complexes at cell membranes. These stable complexes are likely further fortified by additional recruitment of membrane-standing and cytoplasmic PCP components to refine and stabilize correct polarity of epidermal cells and to fix tissue-level PCP information. A role for morphogenetic movements and mechanical strain was also demonstrated in the *Drosophila* wing (Aigouy et al., 2010; Eaton and

Jülicher, 2011), indicating that this might indeed represent an evolutionarily conserved mechanism to initiate bias in planar polarized systems.

Establishment of Planar Cell Polarity in Multiciliated Cells of the Xenopus Epidermis

The process of MCC specification and cilia formation is especially well worked out in the *Xenopus* system (Walentek and Quigley, 2017). Not only do we know in impressive detail which signaling, transcriptional, and post-transcriptional mechanisms regulate MCC formation and ciliogenesis, but the superficial localization and large cell size of MCCs in *Xenopus* also allowed researchers to follow the molecular and morphogenetic steps that MCCs undergo before, during, and after cilia formation (Werner and Mitchell, 2012). This was indeed very informative for our understanding of how these cells establish correct PCP alignment at the cellular level in the context of muco-ciliary epithelia.

The mucociliary epidermis is derived from two layers of epidermal precursor cells, which are generated through asymmetric cell division of animal cells in the *Xenopus* blastula (Huang and Niehrs, 2014). Initially, apicobasal asymmetric cell division produces outer and inner layer cells and leads to unequal distribution of the canonical Wnt/β-catenin coreceptor LRP6 (Fig. 7.2A) (Huang and Niehrs, 2014). High levels of LRP6 in deep layer cells sensitize these cells to canonical Wnt signaling, which is thought to influence cell identity. Outer cells eventually form goblet-like cells, whereas all other epithelial cell types are generated from deep layer cells, which express markers similar to basal airway stem cells (e.g., Tp63, which is a direct target of Wnt/β-catenin) and could potentially represent basal-like cells (Kjolby and Harland, 2016; Lu et al., 2001; Rock et al., 2009; Walentek and Quigley, 2017). The asymmetric distribution of LRP6 among daughter cells is achieved by basolateral localization of LRP6 in ectodermal precursors. Basolateral LRP6 localization requires input from the partitioning-defective (PAR) and Wnt/PCP signaling pathways (Huang and Niehrs, 2014). The PAR pathway is a central regulator of apicobasal polarity in many systems (Goldstein and Macara, 2007; St Johnston and Ahringer, 2010). In *Xenopus* epidermal precursors, the atypical protein kinase C localizes apically and excludes PAR1 localization from this domain, leading to its basolateral accumulation (Fig. 7.2A) (Ossipova et al., 2007; Werner et al., 2014). Additionally, Wnt/PCP signaling manipulations cause defects in LRP6 distribution among outer and deep cells (Huang and Niehrs, 2014). However, it remains unresolved how PAR and Wnt/PCP signaling pathways interact at the molecular level. Since Wnt/PCP signaling influences Dvl recruitment to membranes and its subcellular behavior, it is conceivable that Wnt/PCP signaling plays a permissive role in this context (Gao and Chen, 2010). This idea is supported by findings that revealed a function for PAR signaling in centriole positioning during directional cell division as well as the fact that Dvl, among other Wnt pathway components, can localize to centrioles (Cervenka et al., 2016; Goldstein and Macara, 2007; Kikuchi et al., 2010; Tabler et al., 2010). Therefore, it is possible that PAR signaling defines the axis of cell division through positioning of Dvl-decorated centrioles, and Wnt/PCP signaling then facilitates recruitment of Dvl to basolateral membranes. In any case, Dvl eventually localizes to basolateral membranes of blastula cells and is proposed to recruit LRP6 to these membranes via direct interactions through specific Dvl domains (PDZ and DEP) with the intracellular domain of LRP6 (Huang and Niehrs, 2014).

After induction of cell fate by sets of cell type-specific transcription factors, deep cells migrate in the apical direction and intercalate in between the epithelial layer of outer cells (Stubbs et al., 2006). This process of radial intercalation applies to MCCs, as well as ionocytes, and SSCs (Fig. 7.2C) (Dubaissi et al., 2014; Quigley et al., 2011; Stubbs et al., 2006; Walentek et al., 2014). Apical intercalation requires correct apicobasal polarity, which is again dependent on PAR and Wnt/PCP signaling (Ossipova et al., 2007, 2015; Werner et al., 2014). Cooperation of these pathways leads to apically directed microtubule outgrowth, which relies on the microtubule-binding and -stabilizing protein Clamp (Fig. 7.2C) (Werner et al., 2014). Since Clamp is also part of the ciliary BB and rootlet, centriolar/BB components (e.g., Centrin4) as well as core PCP proteins could also, in principle, be transported apically during intercalation of MCCs (Ossipova et al., 2015; Werner et al., 2014), although it remains elusive if those structures represent bona fide BBs or other types of protein complexes. Importantly, the process of apical surface expansion during epithelial insertion of deep layer cells requires coordinated reorganization of the apical cortical F-actin network, and it was shown that formation of this network relies on Dvl in MCCs (Park et al., 2008; Sedzinski et al., 2016). As mentioned above, Dvl associates with BBs and centrioles; furthermore, it is known to regulate F-actin dynamics in cells through various noncanonical Wnt pathways, including Wnt/PCP. Within these signaling cascades, Dvl-dependent activation of downstream pathway components, including small GTPases (e.g., RhoA, Rock), myosin II regulatory light chain, and Cdc42 (Semenov et al., 2007), can directly influence F-actin behavior and, thus, provide a potential reason for the requirement of Wnt/PCP in this process.

After successful intercalation, MCCs and other cells from the deep ectodermal layer need to synchronize their PCP alignment with cells from the outer layer, i.e., goblet-like cells. Recent work by Butler et al. has investigated the temporal dynamics of core PCP component localization within epidermal cells in *Xenopus* (Butler and Wallingford, 2015). Using overexpressed tagged proteins, symmetrical localization of Fzd6, Dvl1, Van Gogh−like 1 (Vangl1), and the LIM domain−containing intracellular component Prickle2 (Pk2) was detected immediately after the phase of MCC intercalation. However, these components subsequently sorted into distinct asymmetric complexes comprised of Fzd6 and Dvl1 at anterior membranes and Vangl1 and Pk2 at posterior membranes. Although strong asymmetries of these core PCP proteins along the AP axis were observed in both goblet-like cells and MCCs simultaneously, it is most likely that goblet-like cells were already polarized through epidermal stretching at that time, which facilitates noncell autonomous transmission of PCP cues from goblet-like cells to newly intercalated cells from the deep layer (Fig. 7.2D-I) (Butler and Wallingford, 2015; Chien et al., 2015). This view is further supported by the observation of perturbed orientation of wild-type MCCs after manipulation of core PCP components exclusively in outer layer cells (Mitchell et al., 2009). Noncell autonomous propagation of PCP asymmetry is achieved by heterophilic interactions of Fzd6 and Vangl1 across adjacent cell membranes (Fig. 7.2D-I) (Devenport, 2014). Asymmetric sorting and stabilization of Vangl1 and Dvl1 at opposite membranes is dependent on Pk2 (Butler and Wallingford, 2015). Asymmetric localization of Pk2 requires intracellular interactions of Pk2 with Inversin (Invs) and other Pk molecules, while membrane recruitment and asymmetric localization of Vangl1 depends on the presence of Pk2's LIM domains. Pk molecules were previously shown to promote and stabilize Vangl localization at membrane domains, while preventing Dvl localization to these domains (Fig. 7.2D-I) (Butler and Wallingford, 2015; Yang and Mlodzik, 2015). Therefore, asymmetric posterior localization of Pk2 in MCCs leads to asymmetric recruitment of Vangl1 posteriorly and localization of Dvl1 anteriorly, which is where Dvl1 is stabilized by interactions with Fzd6 and Invs. Vangl1-containing cytoplasmic speckles were observed during the process of asymmetric sorting, suggesting that Vangl1 undergoes transcytosis and directional posterior transport along polarized microtubules in MCCs as it was shown in other systems for a variety of core PCP components (Fig. 7.2D-I) (Butler and Wallingford, 2015; Matis et al., 2014).

In summary, the combination of noncell autonomous heterophilic interactions across membranes together with intracellular Pk2-dependent positive and negative regulation leads to intracellular asymmetric sorting of core PCP complexes along the AP axis, which is further enhanced over time. Ultimately, these processes generate and stabilize PCP alignment of individual MCCs in the mucociliary epidermis, while the initial tissue-level PCP is achieved by formation of the first stable core PCP component clusters at membranes of goblet-like cells induced by mechanical strain during gastrulation.

Mouse Trachea

Mammalian conducting airway epithelia are comprised of a majority of MCCs, and other cells including secretory and basal cells (Rock et al., 2010). Directional and concerted beating of motile cilia is essential for clearance of mucus from the lung (Ibanez-Tallon et al., 2003). For effective mucociliary clearance, motile cilia must be oriented and beat along the proximal–distal (lung–oral) axis. This orientation relies on the establishment of tissue-wide PCP in the trachea during embryogenesis (Vladar et al., 2015). This phenomenon is initiated before MCC differentiation. While the global cue that initially defines the proximal–distal axis is unknown, it is clear that cell–cell coordination along this axis relies on the asymmetric distribution of core PCP proteins at cell membranes. This has been extensively studied by Vladar et al. (2012), who analyzed the subcellular distribution of core PCP components using both, specific antibodies and fusion proteins, in primary cultures of mouse tracheal epithelial cells (MTECs) and in vivo (Fig. 7.3D). In MTECs, Vangl1 and Vangl2, Pk2, Fzd3 and Fzd6, Dvl1 and Dvl3, and Celsr1 (Fmi homolog) segregate asymmetrically to cell–cell junctions (Vladar et al., 2015). Similar asymmetric localization of Vangl1, Vangl2, Pk2, and Fzd6 was observed in vivo. Similar to *Xenopus* ectoderm, Vladar et al. described a planar-polarized microtubule network during establishment of PCP in MTECs and showed that microtubule depolymerization impairs the distribution of PCP proteins (Fig. 7.3A). Analyses of mutant mice revealed that Vangl1 is necessary for the recruitment of Pk2 and Vangl2, and to restrict Celsr1 and Fzd6 to specific membrane domains. Finally, Vangl homologs and Pk2 are enriched on the distal side, whereas Fzd homologs, Dvl1, and Dvl3 accumulate at proximal cell membranes. A similar pattern of distribution of core PCP components has been suggested for mouse ependymal cells (Fig. 7.4D) (Boutin et al., 2014; Guirao et al., 2010). This pattern is identical to the one reported in the *Xenopus* mucociliary epidermis and also fits what is known for fly homologs.

The asymmetric distribution of core PCP proteins arises progressively during tracheal development starting before MCCs differentiation (Fig. 7.3B) (Vladar et al., 2012, 2016). Therefore, like in other multiciliated epithelia, MCCs do not participate in the initial establishment of tissue-level PCP. Interestingly, however, Vladar et al. (2016) recently demonstrated that tracheal MCCs are required to reinforce and to maintain tissue-wide polarity. This is evidenced by specific strengthening of PCP protein–partitioning at MCC membranes. Furthermore, at least one PCP gene (Pk2) is expressed in response to MCC-specific transcription factors (e.g., Foxj1) (Fig. 7.3C). Finally, PCP is also disrupted in human airway epithelial cells from cystic fibrosis patients, which form fewer MCCs, and robust PCP could be partially rescued when MCCs were experimentally induced in these cultures (Vladar et al., 2016). Therefore, induction of MCCs could possibly represent an avenue for the improvement of mucociliary clearance in chronic airway diseases.

Mouse Ependyma

Like in the other models, tissue-level PCP of the brain ventricular walls is set up before MCC differentiation. The earliest-known manifestation of

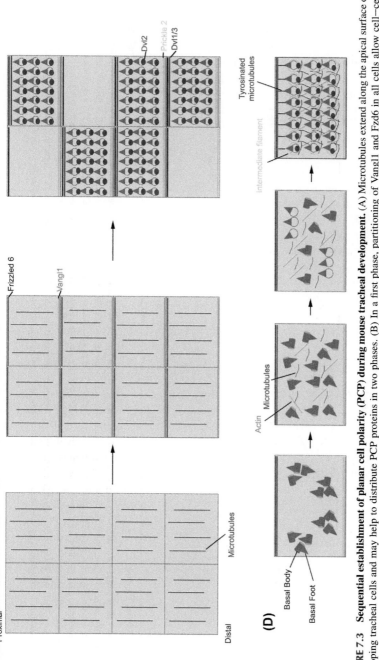

FIGURE 7.3 Sequential establishment of planar cell polarity (PCP) during mouse tracheal development. (A) Microtubules extend along the apical surface of developing tracheal cells and may help to distribute PCP proteins in two phases. (B) In a first phase, partitioning of Vangl1 and Fzd6 in all cells allow cell–cell coordination along the proximal–distal axis of the airway. (C) In a second phase, during MCCs differentiation, PCP proteins partitioning is reinforced specifically in MCCs and two distinct membrane identities emerge: Fzd6 and Dvl1/3 on the proximal side, and Vangl1 and Pk2 on the distal side. (D) Detailed sequence of events at the level of basal bodies (BBs). After docking at the apical membrane, BBs adopt a stereotypic behavior. They first appear as clusters (*first frame*) before spreading (*second frame*) and alignment along rows with similar orientation (*third frame*). This process involves important remodeling of actin, intermediate filament, and microtubule cytoskeletal networks. In mature cells, specific tyrosinated microtubules connect the proximal membrane cortex to the closest row of BBs, ensuring their correct alignment relative to tissue-level polarity (*fourth frame*).

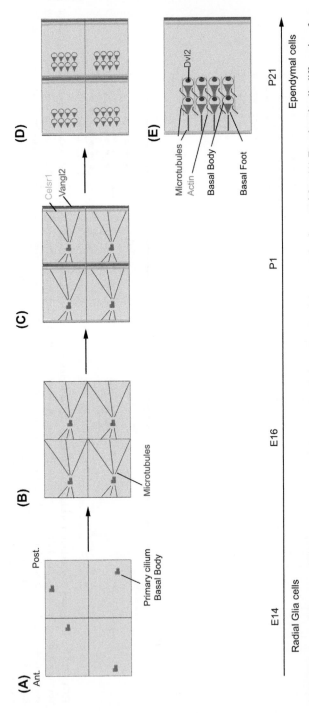

FIGURE 7.4 Sequential establishment of planar cell polarity (PCP) in lateral walls of the mouse brain ventricles. (A) Ependymal cells differentiate from radial glia cells, which initially compose the embryonic walls of mouse brain ventricles. (B) Between embryonic days E14.5 and E16.5, primary cilia of radial glia cells acquire a coordinated polarity. This step functionally involves the PCP proteins Celsr1, Vangl2, and Fzd3. (C) At postnatal day P1, a network of microtubules extends from the primary cilium basal body (BB) toward the cell membrane, which may help to reinforce partitioning of Celsr1 and Vangl2. (D) In mature ependymal cells, BBs appear clustered in a patch placed at a position comparable to that of the primary cilium in radial glia cells. At the tissue scale, positioning and orientation of BB patches is coordinated between cells. (E) The precise organization of BB rows involves local PCP signaling, through Celsr2, Celsr3, Vangl2, and Dvl2, and this is maintained by microtubules and actin filaments. Finally, microtubule filaments connect BBs to the cortical membrane.

polarization concerns radial glia (RG) cells that compose the walls of the developing ventricles, and which are direct precursors of ependymal cells. RGs bear a primary cilium at their apical surface facing the ventricular lumen. During embryogenesis, this cilium is progressively moved toward one side of the apical surface, and this polarization event is locally coordinated between RG cells (Fig. 7.4B). The global cue directing RG cell polarization along this axis is unknown. PKD1 and PKD2 genes, encoding mechanosensory calcium channels, as well as core PCP genes (e.g., Vangl2, Fzd3 and Celsr1) are all necessary for the coordination of primary cilium positioning in RG cells (Boutin et al., 2014; Ohata et al., 2015). However, the cellular mechanisms by which these genes impact on primary cilium localization are still unknown. An attractive hypothesis would be a role for PCP proteins in positioning of the midbody remnant or centrioles following cell division, as they are known to influence primary cilium localization (Bernabé-Rubio et al., 2016). Mutant mice, in which the primary cilium is absent or in which RG polarity is impaired, display defective PCP in the mature ependyma, demonstrating that RG polarity not only prefigures MCC polarization but is also essential for the future PCP of ependymal MCCs (Boutin et al., 2014; Mirzadeh et al., 2010; Ohata et al., 2015). In the developing ependyma, as in the models described above, cell–cell coordination along the axis of polarity relies on planar polarized microtubules, which distribute core PCP proteins asymmetrically. In RG cells, a microtubule network emanating from the BB of the primary cilium and extending toward the cell membrane has been described (Fig. 7.4B) (Boutin et al., 2014). Clear partitioning of the core PCP proteins Vangl2 and Celsr1 has been observed in RG cells after polarization of the primary cilium, and this distribution is abolished when microtubules are depolymerized (Boutin et al., 2014). In mutant mice, in which primary cilium position is not coordinated between cells, Vangl2 remains confined to restricted membrane domains in individual cells, but fails to be coordinately distributed at the tissue level (Boutin et al., 2014). Therefore, the microtubule network emanating from the BB of the primary cilium may contribute to directional sorting of core PCP proteins, which would allow local cell–cell coordination in response to non-cell autonomous cues relaying global tissue polarization to individual cells (Fig. 7.4C). This initial polarity blueprint will be directly transmitted to MCCs during differentiation, to maintain tissue-level PCP (Fig. 7.4D).

TRANSMISSION OF PLANAR CELL POLARITY TO BASAL BODIES IN MULTICILIATED CELLS

After PCP alignment is generated at the global, tissue, and cellular levels, the spatial information influences a diverse set of cell behaviors in a context-dependent manner. These include directional cell migration, asymmetric cell division, convergent extension, and asymmetric genesis, and function of cell organelles or appendages (Wallingford, 2006, 2010; Yang and Mlodzik, 2015).

This is also true for the various types of ciliated cells, including MCCs of mucociliary epithelia. Cilia are generated from modified centrioles called BBs (Fig. 7.1C,E and H). BBs, like centrioles, act as microtubule organizing and polarizing centers in the cell and represent a prerequisite for cilia formation (Marshall, 2008; Zhang and Mitchell, 2016). In MCCs, BB biogenesis occurs through an alternative pathway, which depends on the formation of specialized membraneless organelles, i.e., the deuterosomes. Deuterosomes provide a platform for massive parallel amplification of BBs, which allows MCCs to generate dozens to hundreds of motile cilia at their apical pole (Al Jord et al., 2014; Klos Dehring et al., 2013; Zhang and Mitchell, 2015; Zhao et al., 2013). The hallmark feature of mucociliary epithelia is the generation of directional extracellular fluid flow along their surface (Freund et al., 2012; Marshall and Kintner, 2008). The generation of directional fluid flows along mucociliary epithelia depends on the uniform alignment of MCC BBs as well as their mechanical coupling, which allows coordinated metachronal synchronous beating (Wallingford, 2010). The directional uniform alignment of BBs in MCCs is controlled by the Wnt/PCP pathway. Hence, the question arises how the cellular polarity information set by core PCP components is transmitted to BBs in MCCs? BBs are composed of hundreds of proteins that form structural and enzymatic complexes required for ciliogenesis and directional beating, many of which remain unknown or lack thorough functional characterization. Nevertheless, several studies provide useful information about the processes that lead to BB alignment and generation of directional extracellular fluid flow, which will be discussed in this section.

The *Xenopus* Mucociliary Epidermis

Apical Transport and Membrane Docking of Basal Bodies in Multiciliated Cells

After generation of BBs at deuterosomes in the cytoplasm, these BBs need to be transported to the apical pole of the cell (Avasthi and Marshall, 2012). In contrast to apical transport of core PCP components, which is thought to rely on polarized transport along microtubules (Fig. 7.2C), apical transport of BBs depends on the formation of the actin cytoskeleton (Boisvieux-Ulrich et al., 1990; Dawe et al., 2007). Manipulation of several core PCP proteins as well as downstream effectors of the pathway leads to severe defects in apical transport and impaired cilia formation (Park et al., 2006, 2008; Walentek et al., 2015; Wallingford and Mitchell, 2011), but the mechanisms linking Wnt/PCP signaling to apical BB transport remain incompletely resolved. Interestingly, the PCP effector Inturned (In) requires Vangl proteins for subcellular localization and localizes itself to BBs early on (i.e., while they are still deep in the cytoplasm). In is not only required for apical BB transport but also for the recruitment of other PCP effectors to the BB, including Dishevelled-associated activator of morphogenesis 1 (Daam1)

(Park et al., 2006; Yasunaga et al., 2015). In this process, In interacts with Nephronophthisis protein 4 (NPHP4), which is also required for Daam1 recruitment to BBs (Yasunaga et al., 2015). NPHP4 is known to act in a complex with other NPHP proteins, including NPHP1, which links NPHP4 to Invs (also called NPHP2) (Sang et al., 2011). Invs, in turn, directly interacts with Dvl proteins, and Dvl activates Daam1 (Gao and Chen, 2010; Lienkamp et al., 2012). As mentioned before, Dvl regulates F-actin formation and organization via small GTPases (e.g., RhoA), via Daam1, and via other effectors, such as septins. It is therefore attractive to speculate that in the process of apical BB transport, Wnt/PCP components are primarily required for the promotion of F-actin assembly and organization, but drawing such a conclusion is certainly premature at this point and will require further investigation.

Additionally, proteins from the focal adhesion complex as well as the centrosomal protein with 110 kDa (Cp110) are involved in apical BB transport (Antoniades et al., 2014, 2016; Walentek et al., 2016). Recent studies have demonstrated that focal adhesion kinase (FAK), vinculin, paxillin, and palin are not only localized to focal adhesions but are also found at BBs. There, these factors form ciliary adhesion complexes, which are required for BB interactions with F-actin, and loss of FAK leads to defects in apical BB transport and ciliogenesis (Antoniades et al., 2014). Cp110 is a central protein in centrioles and BBs. Cp110 does not have enzymatic functions by itself, but acts as scaffolding factor to recruit various protein complexes to centrioles and BBs (Song et al., 2014; Tsang and Dynlacht, 2013; Walentek et al., 2016). In MCCs, Cp110 already localizes to BBs while they are still in the deep cytoplasm, and knockdown of *cp110* prevents ciliary adhesion complex formation, apical transport of BBs, and ciliogenesis (Walentek et al., 2016). Thus, it appears conceivable that Cp110 facilitates ciliary adhesion complex formation at BBs, which can connect BBs to F-actin by providing F-actin polymerization and interaction foci. Once F-actin polymerization is initiated and BBs are linked to the actin cytoskeleton, Wnt/PCP components could promote F-actin outgrowth, organization, and directional transport of BBs, possibly in concert with the PAR pathway (Fig. 7.2C).

After BBs are positioned close to the apical membrane, they need to dock to the membrane to allow for ciliogenesis. This process requires fusion of the BB with a ciliary vesicle, derived from the trans-Golgi network (Taschner et al., 2011; Werner and Mitchell, 2012). Ciliary vesicles can attach to the distal end of BBs through interaction with subdistal appendages generated during BB maturation, e.g., by decoration of the BB with Ninein (Graser et al., 2007; Yee and Reiter, 2015). Loss of Dvl2 causes defects in apical BB transport and docking, but analysis of transmission electron micrographs indicates that vesicles are still present close to the apical membrane (Park et al., 2008). This finding suggests that ciliary vesicles are transported to the apical membrane independently of BBs (possibly via microtubules), and that spatial proximity of BBs and ciliary vesicles close to the apical membrane promotes their fusion. Therefore, defective membrane docking of BBs after manipulation of the Wnt/PCP pathway might represent a secondary effect.

Alignment of Basal Bodies and Generation of Unidirectional Extracellular Fluid Flow

Finally, after successful apical transport, ciliary vesicle fusion, and attachment to the apical membrane, BBs start growing axonemes, which need to be aligned in regard to the AP axis in MCCs. Therefore, the cellular directional information generated by core PCP components at lateral membranes needs to be passed on to the many individual BBs found in MCCs (Wallingford, 2010). A series of elegant studies, mainly from Mitchell et al., has convincingly demonstrated that this task is executed in two steps: (1) Initial transmission of PCP information to BBs via an apical microtubule meshwork (Fig. 7.2D-II), and (2) flow-dependent refinement of BB alignment and establishment of metachronal synchronicity via subapical actin filaments (Fig. 7.2E) (Mitchell et al., 2007; Werner et al., 2011).

Initial alignment of BBs in MCCs can be observed early after BBs start to form cilia. In early MCCs, approximately 30%−60% of BBs show correct AP-directional alignment (Mitchell et al., 2007), and these BBs are predominantly located close to the posterior (Vangl1-positive) membrane (Fig. 7.2D-II) (our unpublished observation, Walentek). Interference with microtubule formation during early stages of MCC ciliogenesis, i.e., when the apical microtubule network is formed, completely prevents alignment of BBs. In contrast, inhibition of F-actin formation during these stages prevents correct spacing of BBs and establishment of uniform alignment at the cellular level, but groups of adjacent BBs are still able to align themselves in local patches (Werner et al., 2011). This suggests that BB alignment is mainly achieved by the apical microtubule network, which connects to BBs via their basal feet (Fig. 7.2D-II) (Clare et al., 2014; Werner et al., 2011). This is in line with findings in mammalian systems, where loss of the basal foot does not interfere with cilia formation, but alignment of BBs and generation of directional flow (Kunimoto et al., 2012). Furthermore, the observation of initial alignment of BBs preferentially close to the posterior side of MCCs, i.e., the side toward which basal feet point in the *Xenopus* epidermis, indicates microtubule-dependent interactions between basal feet of BBs and core PCP components at posterior membranes (Fig. 7.2D-II), which was proposed in the mouse (Vladar et al., 2012). However, it remains elusive which basal foot proteins and which PCP components are directly involved in this process.

After a subset of BBs is aligned, cilia beating generates an extracellular fluid flow that is relatively weak and only partially aligned at the beginning. Nevertheless, this initial fluid flow is directionally biased enough to serve as a cue for further alignment of the remaining BBs within individual MCCs and across the whole epidermis (Mitchell et al., 2007). In addition to the apical microtubule network, F-actin filaments are formed subapically, which connect the tip of the striated rootlet from one BB, with the tip of the basal foot from an adjacent BB (Fig. 7.2E) (Antoniades et al., 2014; Werner et al., 2011). This subapical actin network is thought to account for mechanical coupling between BBs. Thereby, it contributes to refinement of BB alignment and provides a basis for metachronal synchronous cilia beating. Like apical BB transport, subapical actin formation depends on the presence of Cp110 and ciliary adhesion complexes, which accumulate at the rootlet tip as well as adjacent to the BB in a posteriorly polarized fashion (Fig. 7.2E) (Antoniades et al., 2014; Walentek et al., 2016). Although it remains to be tested, the posterior localization of Cp110 and ciliary adhesion complexes at BBs suggests that they might form at the tip of the basal foot. This hypothesis is supported by functional studies of atypical tubulins in MCCs, especially zeta-tubulin. Zeta-tubulin forms a complex with other atypical tubulins at the tip of the basal foot (Turk et al., 2015). Loss of zeta-tubulin in *Xenopus* MCCs leads to impaired BB spacing, loss of subapical actin formation, and loss of cell-wide planar polarization of BBs. Importantly, the formation of apical microtubules is not inhibited by loss of zeta-tubulin, and groups of adjacent BBs are still capable to align locally (Turk et al., 2015). These findings distinguish zeta-tubulin from gamma-tubulin, which is also found at basal feet (Clare et al., 2014; Hagiwara et al., 2000). Collectively, these data indicate that zeta-tubulin and other atypical tubulins do not serve as microtubule-nucleating centers as described for gamma-tubulin, but rater provide a structural basis for interaction with F-actin networks. Therefore, the basal foot emerges as a crucial site of BB polarization by formation of complexes that can interact with both the apical microtubule network and the subapical actin network. At the molecular level, these interactions would be established through a gamma-tubulin−containing and microtubule polymerization−promoting complex and a zeta-tubulin/Cp110/ciliary adhesion-containing and F-actin polymerization−promoting complex. It is also likely that several PCP proteins contribute to the establishment of this "posterior polarizing complex" (Fig. 7.2E), since Dvl2 and Pk2 were found to localize adjacent to BBs in a polarized fashion as well and knockdown of the small GTPase Rac1 caused essentially the same MCC phenotype as loss of zeta-tubulin, or Cp110, or FAK (Antoniades et al., 2014; Butler and Wallingford, 2015; Epting et al., 2015; Park et al., 2008; Walentek et al., 2016). The localization of PCP components at such "posterior polarizing complex" could also provide the mechanistic basis for PCP-dependent regulation of F-actin formation and organization of MCC BBs as well as PCP-dependent initial polarization of BBs near the posterior membrane of MCCs: On the one hand, F-actin could be regulated by the locally restricted function of PCP effectors, such as Daam1, small GTPases, and septins; on the other hand, Pk2 could directly interact with membrane-associated Pk molecules, which form stable complexes exclusively at the posterior membrane in *Xenopus* MCCs.

Taken together, these observations indicate there are several time points and sites within the cell, when and where the Wnt/PCP pathway controls BB behavior in MCCs. First, PCP proteins control apical F-actin–dependent transport of BBs, and then they are required for initial basal foot- and microtubule-dependent alignment close to the posterior pole of MCCs, and lastly, PCP contributes to flow- and subapical actin-dependent refined and uniform alignment of BBs as well as metachronal synchronous ciliary beating. In these processes, specific interactions of the BB with the cytoskeleton are required, which are mediated by gamma-tubulin for microtubules, and zeta-tubulin, Cp110, and ciliary adhesions for F-actin networks. It will be very interesting to further address the many open questions regarding the mechanistic details regulating these processes. Especially, it might be very informative to test the hypothesis that PCP components, ciliary adhesions, and gamma-/zeta-tubulin all form a "posterior polarizing complex" at the tip of the ciliary foot, which could integrate and orchestrate PCP information with cytoskeletal organization, and thereby establish coordinated directional beating of cilia in MCCs of mucociliary epithelia.

Mouse Trachea

In tracheal MCCs, BBs are generated through centrosome and deuterosome pathways and are released in the subapical space before migrating to the cell surface (Zhao et al., 2013). Early electron microscopy studies in the rat embryonic lung suggested that apical BB docking is preceded by recruitment of ciliary vesicles to distal appendages of BBs (Sorokin, 1968). A recent study has shown that this process is dependent on recruitment of the coiled-coil protein Chibby (Cby) during centriole amplification (Burke et al., 2014). Cby is recruited to the distal end of growing centrioles by physical interaction with the protein Cep164. By interacting with the Rab8 GEF Rabin8, Cby enhances the recruitment of Rab8 positive ciliary vesicles that fuse with distal ends of BBs and enable correct apical membrane docking of BBs.

Once BBs have attached at the apical membrane, they adopt a typical distribution and polarized orientation. In adult mouse MCCs and cultured MTECs, BBs are arranged in parallel rows that cover the entire apical surface of the cell with a regular spacing between rows (Fig. 7.3C). This distribution contrasts with that of *Xenopus* MCCs in which BBs are evenly spaced and are not organized in rows. In tracheal MCCs, analysis of BB appendages indicates that each row of BBs is orientated in the same direction, which ensures co-ordinated ciliary beating at the single cell level (Fig. 7.3C–D). Combining long-term high-resolution live-cell imaging of MTECs and analysis of fixed samples, a recent study described the emergence of this specific organization of the BB array (Herawati et al., 2016). This study showed that BBs first appear in small clusters and subsequently spread before progressively aligning to reach their final position (Fig. 7.3C). These movements of BBs coincide with the progressive coordination of their orientation, suggesting that these two processes—alignment and orientation of BBs—may share molecular regulators and/or be partially dependent on each other.

Similar to other MCCs, MTECs display a dense network of intermingled cytoskeletal elements just below the apical membrane. In mature MTECs, this apical meshwork is composed of (1) actin filaments occasionally connected with BBs and basal feet, (2) Keratin 8 positive intermediate filaments that encircle each BB, and (3) microtubules distributed parallel to the apical surface and associated with basal feet through components such as ODF2 (Herawati et al., 2016; Kunimoto et al., 2012). Actin is involved in BBs spacing and alignment, whereas microtubules are important for both alignment and orientation of BBs (Fig. 7.3C). Interestingly, the molecular cross-linker Plectin, which can mediate interactions between microtubules, intermediate filaments, and actin fibers, is observed in close vicinity to intermediate filaments. This suggests that in MTECs, the three types of cytoskeleton may interact with each other and cooperate to control spacing, alignment, and orientation of the BB array (Herawati et al., 2016).

In tracheal MCCs, like in other multiciliated epithelia, cilia must align with the axis of tissue polarity to ensure coordinated ciliary beating and generation of extracellular fluid flow at the epithelial lining of the organ. This implies that tracheal MCCs have to detect and interpret PCP information of adjacent membranes to correctly align their BBs. In tracheal cells, a specific network of tyrosinated microtubules connects the basal feet of the first row of BBs to the proximal Fzd/Dvl positive membrane domain (Fig. 7.3C). This has been proposed to instruct correct orientation to the BB array. Vladar et al. (2012) further suggested that these microtubules are necessary for interpretation of tissue-level PCP, independently of the apical microtubule network connecting BBs within the array. Although not formally demonstrated, this hypothesis is strongly supported by the analysis of trachea of Vangl1 mutant in which tissue-level polarity is impaired, with little impact on internal BB array orientation.

Mouse Ependyma

The organization of BBs at the apical surface of mature ependymal MCCs strikingly differs from that in all other known MCCs. They are not distributed uniformly throughout the entire surface but are confined in a well-defined patch that covers only about 20% of the apical membrane. How is the ciliary patch established and maintained is not known but represents a matter of great interest. The BB patch is positioned in the anterior part of the cell, similarly to the primary cilium in the RG progenitor (Mirzadeh et al., 2010). However, the functional relevance of this particular localization of the ciliary tuft has not been tested. Within a patch, BBs are distributed and oriented in a very stereotypical manner, comparable, although not strictly identical, to the organization observed in *Xenopus* epidermal and mouse tracheal MCCs. BBs are organized in parallel rows with uniformly oriented basal feet. The distance between each row is larger than the distance between BBs that comprise a given row (Fig. 7.4E) (Boutin et al., 2014; Mirzadeh et al., 2010). Therefore, three architectural features must be considered in ependymal MCCs: (1) clustering of BBs in a patch; (2) patch positioning within the cell; and (3) BB organization inside the patch. The vectors defining patch position within the cell and BBs orientation within the patch are aligned with each other and are coordinated between adjacent cells, suggesting that tissue-level PCP organizes both features.

Before discussing how this may be achieved, we will first focus on what molecular and cellular mechanisms regulate the clustering of BBs in a patch and their correct organization within this patch. A way to understand organization is to follow BBs from their place of birth to their final position: The differentiation of ependymal MCCs starts with the initiation of the MCC-specific transcriptional cascade in RG cells (Kyrousi et al., 2015; Meunier and Azimzadeh, 2016; Tan et al., 2013). Similar to the above-described models, amplification of BBs occurs around deuterosomes, which themselves bud off the daughter centriole of RG cells (Al Jord et al., 2014). In ependymal MCCs, the process of amplification is synchronized, such that all centrioles are released from deuterosomes and arrive at the subapical space at the same time. At this stage, they appear broadly distributed at low density (Al Jord et al., 2014; Hirota et al., 2010). From there, in a schematic manner, they need to carry out at least two distinct movements: apical docking and clustering into a patch. Two scenarios can be envisaged that will need to be tested with high-resolution live imaging: (1) Either BBs could first dock in a random manner to the apical membrane before clustering through a translational movement; (2) Or, BB apical migration could be spatially restricted to target a predefined area of the apical surface. Discriminating between the two scenarios would certainly provide a hint on how ependymal cells organize their apical surface in response to tissue-level polarity cues.

Several recent publications have started to elucidate the mechanisms controlling BB organization in ependymal MCCs, revealing important contributions of cytoskeletal elements and PCP proteins (Fig. 7.4E). Sequential remodeling of the apical actin cytoskeleton is an important feature of ependymal cell maturation. These include the emergence of two different actin networks (similar to those described by the Mitchell lab in *Xenopus* epidermal MCCs; Werner et al., 2011). The apical F-actin network appears simultaneously with BB clustering and remains present in mature ependymal MCCs. The second network appears later during ependymal maturation, localizes subapically, and establishes short connections between neighboring BBs (Boutin et al., 2014). Three mammalian nonmuscle myosin IIs (Myh9, Myh10, and Myh14) are expressed in ependymal cells, Myh9 being distributed in the cytoplasm, whereas Myh10 and Myh14 are found closer to BBs (in particular in their phosphorylated active form). Chemical inhibition of myosin II or myosin light chain kinase as well as knockdown of MyH9, Myh10, and Myh14 perturbs BB clustering and anterior positioning (Hirota et al., 2010). In Celsr2 mutant mice, BBs are clustered in a patch with an almost normal size, but within this patch, BB spacing is impaired and the actin meshwork is malformed (Boutin et al., 2014). Taken together, this suggests that the actin cytoskeleton is necessary for two aspects of ependymal BB organization: (1) Actin−myosin cooperation is important for BB clustering, and (2), by analogy to findings from *Xenopus* MCCs, it appears that actin is necessary for the correct spacing of BBs within the patch.

Important microtubule remodeling is also observed at the apical surface of maturing ependymal MCCs and, in fully mature cells, a set of microtubules is intermingled with the actin cytoskeleton at the level of the BB patch (Fig. 7.4E). Although a formal functional proof has not been provided yet, one could speculate that similar to what is observed in *Xenopus* ectodermal MCCs and in MTECs (Herawati et al., 2016; Werner et al., 2011), these microtubules may be important for both alignment and orientation of BBs.

Several studies implicated molecular actors in the organization of BBs. These include members of the core PCP gene family (Celsr2, Celsr3, Fzd3, Dvl2, Vangl2) that control both the spacing (Celsr2) and orientation (Celsr3, Fzd3, Dvl2, Vangl2) of BBs within patches. Interestingly, Celsr2, Celsr3, and Dvl2 specifically impact BB organization inside the patch without affecting tissue-level polarity. Although not firmly established, this subset of PCP components is likely to influence the local organization of the actin and microtubule cytoskeletal networks. In line with this, Dvl2 has been shown to localize to the BBs of ependymal cells (Hirota et al., 2010). These studies revealed a significant degree of specialization of core PCP components, whereby distinct members organize different features of polarization in the ependymal epithelium.

Beside the implication of PCP genes, precise qualitative and quantitative analysis of mutant mice led to intriguing observations: For instance, it is noticeable, that the BB orientation in Celsr3 mutants is not fully randomized. Instead, local BB organization was maintained, generating perfectly aligned and oriented rows but with opposite orientation within a patch (Boutin et al., 2014). This observation suggests that BBs in ependymal MCCs have a certain capacity for local self-organization, independently of global MCC polarity, as it was shown for *Xenopus* MCCs (Werner et al., 2011).

In addition to "local PCP signaling" responsible for organization of BBs inside the patch, the set of "tissue-level PCP genes" is also important to organize the ependymal apical surface (Boutin et al., 2014; Ohata et al., 2014). Within this subset of PCP proteins, Celsr1 localizes uniformly around the cell cortex, whereas Vanlg2 localizes to the posterior cell membrane, away from the BB patch. Such subcellular localization suggests that these proteins may provide information for building intracellular scaffolds, which might help to correctly position and orient the BB patch relative to tissue-level polarity. In agreement with this hypothesis, in Celsr1 mutant mice, patch position and BB orientation in individual cells are incorrectly aligned relative to tissue-level PCP, while intracellularly, patch positioning relative to the Vangl2-positive membrane is preserved. This may reflect the fact that despite a loss of cell–cell coordination, each individual ependymal cell is able to interpret its own membrane-associated PCP information and to position its patch of BBs in a cell autonomous manner. The transmission of polarity information from the membrane to the patch is likely to occur via microtubules connecting the cell cortex to BBs (Fig. 7.4E). However, in this context, the direction of patch displacement and its orientation, which are normally aligned, are uncoupled and sometimes even opposite (a phenotype shared by all tissue-level PCP mutants). This unexpected observation suggests that a lot remains to be uncovered about the mechanisms that allow the transmission of tissue-level polarity to ependymal BBs.

FUTURE CHALLENGES

Outstanding questions that should be addressed to expand and deepen our knowledge of ciliated epithelia organization are highlighted in Boxes 7.1 and 7.2. Here, we would like to briefly point out which challenges must be faced to successfully answer these questions. The main challenge will be to

---

BOX 7.1 Conclusions and Open Questions—Establishment of Planar Cell Polarity (PCP) in Ciliated Tissues

A common scenario seems to emerge that may explain how tissue-level polarity arises in ciliated epithelia. First, a global axis of polarity needs to be established during embryogenesis. The cues responsible for this founding event may be morphogenetic movements in response to expressed Wnt ligands or other cues, which then causes mechanical strain on precursor cells. These global polarity cues are subsequently translated into coordinated alignment of apical microtubules, which allow directional transport of core PCP molecules and the formation of membrane domains containing stabilized complexes of core PCP proteins. This initially weak pattern of asymmetry is reinforced over time through noncell autonomous and cell autonomous self-organization of PCP protein complexes. All these events occur prior to terminal differentiation of MCCs, although MCCs might reinforce global tissue-level polarity secondarily. Several open questions remain to be addressed in the future, including the potential role for the Ft–Ds pathway in vertebrate multiciliated epithelia. Furthermore, it will be important to elucidate the molecular basis for the cross talk between the Wnt/PCP, the Ft–Ds, and the PAR signaling pathways in mucociliary development, regeneration, and disease.

BOX 7.2 Conclusions and Open Questions—Orientation and Alignment of Basal Bodies and Cilia

Comparing our three models of ciliated epithelia, it is clear that actin and microtubule networks are critical for the proper orientation, spacing, and physical coupling of BBs, and ultimately, for the production of robust directional ciliary beating. A key question in this area concerns the molecular nature of the links between polarized MCC membrane domains and BBs. It is possible, that in all models, only some of BBs may interact with the cell membrane to read preset polarity and transmit it to inner BBs. In that respect, it may be useful to think of the ciliated apical surface as a self-organizing raft that needs to anchor at one pole of the MCC lateral membrane.

Since Wnt/PCP signaling regulates actin behavior through various PCP effector proteins, it will be crucial to elucidate how these are recruited and regulated at specific domains of the membrane or at BBs. This information will be important to understand how BBs contribute to the organization and function of cytoskeletal networks during apical BB transport, PCP signaling, BB alignment, and metachronal synchronous cilia beating at the apical membrane of mature MCCs.

Another intriguing conundrum emerges from the current literature: in mouse tracheal and ependymal MCCs, cilia beat away from the Vangl-positive membrane domain, whereas cilia beat toward the Vangl-positive domain in *Xenopus* epidermal MCCs. This apparent paradoxon may reveal the existence of additional polarization factors, which may control cilia orientation independently or coordinately with Wnt/PCP. Alternatively, it may reflect differences in the molecular composition of polarized membranes between models that would change their capacity to interact with BBs and the associated cytoskeleton. Resolving this open question may bring essential new insights to the cell polarity field.

Lastly, it is worth stressing the systematic implication of PCP proteins in all aspects of ciliated epithelia organization, from the tissue level, to the individual cell level, down to BB migration, docking, and orientation. The current evidence reveals a significant level of specialization of individual PCP proteins that has been made possible through gene family expansion (e.g., Dvl2 association to BBs but not cell membranes, and in contrast to other Dvl family members). Therefore, it seems important to improve the characterization of the individual endogenous PCP protein distribution in the various models. Furthermore, it will be important to analyze in depth the individual contributions of family members that may have been overlooked in many cases.

document with very high spatial and temporal resolution the distribution of key polarity effectors of the PCP pathway and of actin, microtubule, and intermediate filament cytoskeletal constituents. It is expected that recent developments in super-resolution fluorescence microscopy and correlative light/electron microscopy will help us to reach this objective. As described above, information about the distribution of specific PCP members is often missing; here, CRISPR−Cas9 mediated knock-in of fluorescent tags in endogenous genes offers great promise to fill this gap. Of course, highly specific antibodies also represent invaluable resources that should be further expanded. Another challenge will consist in inactivating selected factors in a spatiotemporally controlled manner. The task is daunting when thinking about proteins that can be important for PCP at multiple sites in the cell, or at multiple scales of organization, as described in this chapter. Conditional genetic ablation may only bring a partial response to such difficulties, and the use of inducible-mutated constructs targeted at specific subcellular locations also represents a valuable, albeit complex, approach. Hence, it is possible that mathematical modeling will be of great help to identify relevant parameters, when experimental capacity will become limiting. Finally, it may be worth expanding the repertoire of ciliated epithelia to nonmodel organisms that may offer unique advantages to decipher relevant cellular mechanisms (e.g., Rompolas et al., 2009).

GENERAL CONCLUSION

In this chapter, we summarize the current state of the field and highlight why ciliated epithelia are well suited to study PCP across organizational scales, from tissue level to organelle level. The field has contributed to and benefited from the broader field of PCP and will continue to do so. We also wanted to illustrate the importance of studying multiple tissues and organisms to reveal conserved principles as well as divergent modes of organization. Ciliated epithelia play essential roles in human physiology, and altered polarity is expected to cause dysfunctions or contribute to human diseases. However, the molecules and pathways discussed in this review are also essential for embryonic development, which may explain why they did not attract extensive attention of clinicians so far. This situation may change with the advent of unbiased genome-wide clinical analyses, and the knowledge gained through work in model organisms will become invaluable to elucidate the pathophysiological mechanisms underlying human diseases.

ACKNOWLEDGMENTS

The authors thank Rui Song and Lin He for permission to use SEM and TEM images of mouse tracheal MCCs. Peter Walentek is funded by a Pathway to Independence Award by the NIH-NHLBI (K99HL127275) and thanks Richard M. Harland and the Harland lab for support and discussions. Research in L. Kodjabachian's laboratory is supported by CNRS and Aix-Marseille Université and by grants from Agence Nationale de la Recherche, Fondation pour la Recherche Médicale, and Fondation ARC. C. Boutin is supported by a fellowship from Ligue Nationale contre le Cancer. The authors apologize to all researchers whose work was not cited due to space restrictions.

REFERENCES

Aigouy, B., Farhadifar, R., Staple, D.B., Sagner, A., Röper, J.-C., Jülicher, F., Eaton, S., Ju, F., Farhadifar, R., Staple, D.B., et al., 2010. Cell flow reorients the Axis of planar polarity in the wing epithelium of *Drosophila*. Cell 142, 773–786.

Al Jord, A., Lemaître, A.-I., Delgehyr, N., Faucourt, M., Spassky, N., Meunier, A., 2014. Centriole amplification by mother and daughter centrioles differs in multiciliated cells. Nature 516, 104–107.

Andre, P., Song, H., Kim, W., Kispert, A., Yang, Y., 2015. Wnt5a and Wnt11 regulate mammalian anterior-posterior axis elongation. Development 1–12.

Antoniades, I., Stylianou, P., Skourides, P.A., 2014. Making the connection: ciliary adhesion complexes anchor basal bodies to the actin cytoskeleton. Dev. Cell 28, 70–80.

Antoniades, I., Stylianou, P., Christodoulou, N., Skourides, P.A., 2016. Addressing the functional determinants of FAK during ciliogenesis in multiciliated cells.

Avasthi, P., Marshall, W.F., 2012. Stages of ciliogenesis and regulation of ciliary length. Differentiation 83, S30–S42.

Bernabé-Rubio, M., Andrés, G., Casares-Arias, J., Fernández-Barrera, J., Rangel, L., Reglero-Real, N., Gershlick, D.C., Fernández, J.J., Millán, J., Correas, I., et al., 2016. Novel role for the midbody in primary ciliogenesis by polarized epithelial cells. J. Cell Biol. 214, 259–273.

Blum, M., Feistel, K., Thumberger, T., Schweickert, A., 2014. The evolution and conservation of left-right patterning mechanisms. Development 141, 1603–1613.

Boisvieux-Ulrich, E., Lainé, M.C., Sandoz, D., 1990. Cytochalasin D inhibits basal body migration and ciliary elongation in quail oviduct epithelium. Cell Tissue Res. 259, 443–454.

Boon, M., Wallmeier, J., Ma, L., Loges, N.T., Jaspers, M., Olbrich, H., Dougherty, G.W., Raidt, J., Werner, C., Amirav, I., et al., 2014. MCIDAS mutations result in a mucociliary clearance disorder with reduced generation of multiple motile cilia. Nat. Commun. 5, 4418.

Boutin, C., Diestel, S., Desoeuvre, A., Tiveron, M.-C., Cremer, H., 2008. Efficient in vivo electroporation of the postnatal rodent forebrain. PLoS One 3, e1883.

Boutin, C., Labedan, P., Dimidschstein, J., Richard, F., Cremer, H., André, P., Yang, Y., Montcouquiol, M., Goffinet, A.M., Tissir, F., 2014. A dual role for planar cell polarity genes in ciliated cells. Proc. Natl. Acad. Sci. U.S.A. 1–10.

Brooks, E.R., Wallingford, J.B., 2014. Multiciliated cells. Curr. Biol. 24, R973–R982.

Brown, N.A., Wolpert, L., 1990. The development of handedness in left/right asymmetry. Development 109, 1–9.

Burke, M.C., Li, F.Q., Cyge, B., Arashiro, T., Brechbuhl, H.M., Chen, X., Siller, S.S., Weiss, M.A., O'Connell, C.B., Love, D., et al., 2014. Chibby promotes ciliary vesicle formation and basal body docking during airway cell differentiation. J. Cell Biol. 207, 123–137.

Butler, M.T., Wallingford, J.B., 2015. Control of vertebrate core PCP protein localization and dynamics by Prickle2. Development 142, 3429–3439.

Cervenka, I., Valnohova, J., Bernatik, O., Harnos, J., Radsetoulal, M., Sedova, K., Hanakova, K., Potesil, D., Sedlackova, M., Salasova, A., et al., 2016. Dishevelled is a NEK2 kinase substrate controlling dynamics of centrosomal linker proteins. Proc. Natl. Acad. Sci. U.S.A. 113, 9304–9309.

Cha, S.-W., Tadjuidje, E., Tao, Q., Wylie, C., Heasman, J., 2008. Wnt5a and Wnt11 interact in a maternal Dkk1-regulated fashion to activate both canonical and non-canonical signaling in *Xenopus* axis formation. Development 3729, 3719–3729.

Chien, Y.-H.H., Keller, R., Kintner, C., Shook, D.R.R., 2015. Mechanical strain determines the axis of planar polarity in ciliated epithelia. Curr. Biol. 25, 2774–2784.

Choksi, S.P., Lauter, G., Swoboda, P., Roy, S., 2014. Switching on cilia: transcriptional networks regulating ciliogenesis. Development 141, 1427–1441.

Clare, D.K., Magescas, J., Piolot, T., Dumoux, M., Vesque, C., Pichard, E., Dang, T., Duvauchelle, B., Poirier, F., Delacour, D., 2014. Basal foot MTOC organizes pillar MTs required for coordination of beating cilia. Nat. Commun. 5, 4888.

Dawe, H.R., Farr, H., Gull, K., 2007. Centriole/basal body morphogenesis and migration during ciliogenesis in animal cells. J. Cell Sci. 120, 7–15.

Devenport, D., 2014. The cell biology of planar cell polarity. J. Cell Biol. 207, 171–179.

Dubaissi, E., Papalopulu, N., 2011. Embryonic frog epidermis: a model for the study of cell-cell interactions in the development of mucociliary disease. Dis. Model. Mech. 192, 179–192.

Dubaissi, E., Rousseau, K., Lea, R., Soto, X., Nardeosingh, S., Schweickert, A., Amaya, E., Thornton, D.J., Papalopulu, N., 2014. A secretory cell type develops alongside multiciliated cells, ionocytes and goblet cells, and provides a protective, anti-infective function in the frog embryonic mucociliary epidermis. Development 141, 1514–1525.

Eaton, S., Jülicher, F., 2011. Cell flow and tissue polarity patterns. Curr. Opin. Genet. Dev. 21, 747–752.

Epting, D., Slanchev, K., Boehlke, C., Hoff, S., Loges, N.T., Yasunaga, T., Indorf, L., Nestel, S., Lienkamp, S.S., Omran, H., et al., 2015. The Rac1 regulator ELMO controls basal body migration and docking in multiciliated cells through interaction with Ezrin. Development 142, 174–184.

Freund, J.B., Goetz, J.G., Hill, K.L., Vermot, J., 2012. Fluid flows and forces in development: functions, features and biophysical principles. Development 139, 1229–1245.

Fulcher, M.L., Gabriel, S., Burns, K.A., Yankaskas, J.R., Randell, S.H., 2005. Well-differentiated human airway epithelial cell cultures. Methods Mol. Med. 107, 183–206.

Ganesan, S., Comstock, A.T., Sajjan, U.S., 2013. Barrier function of airway tract epithelium. Tissue Barriers 1, e24997.

Gao, C., Chen, Y.-G., 2010. Dishevelled: the hub of Wnt signaling. Cell. Signal 22, 717–727.

Garcia, G., Reiter, J.F., 2016. A primer on the mouse basal body. Cilia 5, 17.

Goldstein, B., Macara, I.G., 2007. The PAR proteins: fundamental players in animal cell polarization. Dev. Cell 13, 609–622.

Goodrich, L.V., Strutt, D., 2011. Principles of planar polarity in animal development. Development 138, 1877–1892.

Graser, S., Stierhof, Y.-D., Lavoie, S.B., Gassner, O.S., Lamla, S., Le Clech, M., Nigg, E.A., 2007. Cep164, a novel centriole appendage protein required for primary cilium formation. J. Cell Biol. 179, 321–330.

Guirao, B., Meunier, A., Mortaud, S., Aguilar, A., Corsi, J.-M., Strehl, L., Hirota, Y., Desoeuvre, A., Boutin, C., Han, Y.-G., et al., 2010. Coupling between hydrodynamic forces and planar cell polarity orients mammalian motile cilia. Nat. Cell Biol. 12, 341–350.

Hagiwara, H., Kano, A., Aoki, T., Ohwada, N., Takata, K., 2000. Localization of gamma-tubulin to the basal foot associated with the basal body extending a cilium. Histochem. J. 32, 669–671.

Hardy, K.M., Garriock, R.J., Yatskievych, T.A., D'Agostino, S.L., Antin, P.B., Krieg, P.A., 2008. Non-canonical Wnt signaling through Wnt5a/b and a novel Wnt11 gene, Wnt11b, regulates cell migration during avian gastrulation. Dev. Biol. 320, 391–401.

Hayes, J.M., Kim, S.K., Abitua, P.B., Park, T.J., Herrington, E.R., Kitayama, A., Grow, M.W., Ueno, N., Wallingford, J.B., Kyoung, S., et al., 2007. Identification of novel ciliogenesis factors using a new in vivo model for mucociliary epithelial development. Dev. Biol. 312, 115–130.

Heasman, J., 2006. Patterning the early *Xenopus* embryo. Development 133, 1205—1217.

Herawati, E., Taniguchi, D., Kanoh, H., Tateishi, K., Ishihara, S., Tsukita, S., 2016. Multiciliated cell basal bodies align in stereotypical patterns coordinated by the apical cytoskeleton. J. Cell Biol. 214, 571—586.

Hikasa, H., Sokol, S.Y., 2013. Wnt signaling in vertebrate axis specification. Cold Spring Harb. Perspect. Biol. 5, 1—20.

Hirota, Y., Meunier, A., Huang, S., 2010. Planar polarity of multiciliated ependymal cells involves the anterior migration of basal bodies regulated by non-muscle myosin II. Development 137, 3037—3046.

Hogan, B.L.M., Barkauskas, C.E., Chapman, H.A., Epstein, J.A., Jain, R., Hsia, C.C.W., Niklason, L., Calle, E., Le, A., Randell, S.H., et al., 2014. Repair and regeneration of the respiratory system: complexity, plasticity, and mechanisms of lung stem cell function. Cell Stem Cell 15, 123—138.

Houston, D.W., 2012. Cortical rotation and messenger RNA localization in *Xenopus* axis formation. Wiley Interdiscip. Rev. Dev. Biol. 1, 371—388.

Huang, Y., Niehrs, C., 2014. Polarized Wnt signaling regulates ectodermal cell fate in *Xenopus*. Dev. Cell 29, 250—257.

Ibanez-Tallon, I., Heintz, N., Omran, H., 2003. To beat or not to beat: roles of cilia in development and disease. Hum. Mol. Genet. 12, 27R—35R.

Kikuchi, K., Niikura, Y., Kitagawa, K., Kikuchi, A., 2010. Dishevelled, a Wnt signalling component, is involved in mitotic progression in cooperation with Plk1. EMBO J. 29, 3470—3483.

Kjolby, R.A.S., Harland, R.M., 2016. Genome-wide identification of Wnt/β-catenin transcriptional targets during *Xenopus* gastrulation. Dev. Biol. 1—11.

Klos Dehring, D.A., Vladar, E.K., Werner, M.E., Mitchell, J.W., Hwang, P., Mitchell, B.J., 2013. Deuterosome-mediated centriole biogenesis. Dev. Cell 27, 103—112.

Kunimoto, K., Yamazaki, Y., Nishida, T., Shinohara, K., Ishikawa, H., Hasegawa, T., Okanoue, T., Hamada, H., Noda, T., Tamura, A., et al., 2012. Coordinated ciliary beating requires Odf2-mediated polarization of basal bodies via basal feet. Cell 148, 189—200.

Kyrousi, C., Arbi, M., Pilz, G.A., Pefani, D.E., Lalioti, M.E., Ninkovic, J., Gotz, M., Lygerou, Z., Taraviras, S., 2015. Mcidas and GemC1 are key regulators for the generation of multiciliated ependymal cells in the adult neurogenic niche. Development 142, 3661—3674.

Le Pabic, P., Ng, C., Schilling, T.F., 2014. Fat-dachsous signaling coordinates cartilage differentiation and polarity during craniofacial development. PLoS Genet. 10.

Lienkamp, S., Ganner, A., Walz, G., 2012. Inversin, Wnt signaling and primary cilia. Differentiation 83, S49—S55.

Lim, D.A., Alvarez-Buylla, A., 2014. Adult neural stem cells stake their ground. Trends Neurosci. 37, 563—571.

Lu, P., Barad, M., Vize, P.D., 2001. *Xenopus* p63 expression in early ectoderm and neurectoderm. Mech. Dev. 102, 275—278.

Mao, Y., Mulvaney, J., Zakaria, S., Yu, T., Morgan, K.M., Allen, S., Basson, M.A., Francis-West, P., Irvine, K.D., 2011. Characterization of a Dchs1 mutant mouse reveals requirements for Dchs1-Fat4 signaling during mammalian development. Development 138, 947—957.

Marshall, W.F., Kintner, C., 2008. Cilia orientation and the fluid mechanics of development. Curr. Opin. Cell Biol. 20, 48—52.

Marshall, W.F., Nonaka, S., 2006. Cilia: tuning in to the cell's antenna. Curr. Biol. 16, R604—R614.

Marshall, W.F., 2008. Basal bodies : platforms for building cilia. Curr. Top. Dev. Biol. 85, 1—22.

Matis, M., Russler-Germain, D.A., Hu, Q., Tomlin, C.J., Axelrod, J.D., 2014. Microtubules provide directional information for core PCP function. Elife 3, e02893.

Meunier, A., Azimzadeh, J., 2016. Multiciliated cells in animals. Cold Spring Harb. Perspect. Biol. 1—22.

Mirzadeh, Z., Han, Y., Soriano-Navarro, M., García-Verdugo, J.M., Alvarez-Buylla, A., 2010. Cilia organize ependymal planar polarity. J. Neurosci. 30, 2600—2610.

Mitchell, B., Jacobs, R., Li, J., Chien, S., Kintner, C., 2007. A positive feedback mechanism governs the polarity and motion of motile cilia. Nature 447, 97—101.

Mitchell, B., Stubbs, J.L., Huisman, F., Taborek, P., Yu, C., Kintner, C., 2009. The PCP pathway instructs the planar orientation of ciliated cells in the *Xenopus* larval skin. Curr. Biol. 19, 924—929.

Ohata, S., Nakatani, J., Herranz-Pérez, V., Cheng, J., Belinson, H., Inubushi, T., Snider, W.D., García-Verdugo, J.M., Wynshaw-Boris, A., Álvarez-Buylla, A., 2014. Loss of dishevelleds disrupts planar polarity in ependymal motile cilia and results in hydrocephalus. Neuron 83, 558—571.

Ohata, S., Herranz-Perez, V., Nakatani, J., Boletta, A., Garcia-Verdugo, J.M., Alvarez-Buylla, A., 2015. Mechanosensory genes Pkd1 and Pkd2 contribute to the planar polarization of brain ventricular epithelium. J. Neurosci. 35, 11153—11168.

Ossipova, O., Tabler, J., Green, J.B.A., Sokol, S.Y., 2007. PAR1 specifies ciliated cells in vertebrate ectoderm downstream of aPKC. Development 134, 4297—4306.

Ossipova, O., Chu, C.-W., Fillatre, J., Brott, B.K., Itoh, K., Sokol, S.Y., 2015. The involvement of PCP proteins in radial cell intercalations during *Xenopus* embryonic development. Dev. Biol. 11, 1—12.

Park, T.J., Haigo, S.L., Wallingford, J.B., 2006. Ciliogenesis defects in embryos lacking inturned or fuzzy function are associated with failure of planar cell polarity and Hedgehog signaling. Nat. Genet. 38, 303—311.

Park, T.J., Mitchell, B.J., Abitua, P.B., Kintner, C., Wallingford, J.B., 2008. Dishevelled controls apical docking and planar polarization of basal bodies in ciliated epithelial cells. Nat. Genet. 40, 871—879.

Quigley, I.K., Stubbs, J.L., Kintner, C., 2011. Specification of ion transport cells in the *Xenopus* larval skin. Development 714, 705—714.

Rock, J.R., Onaitis, M.W., Rawlins, E.L., Lu, Y., Clark, C.P., Xue, Y., Randell, S.H., Hogan, B.L.M., 2009. Basal cells as stem cells of the mouse trachea and human airway epithelium. Proc. Natl. Acad. Sci. U.S.A. 106, 12771—12775.

Rock, J.R., Randell, S.H., Hogan, B.L.M., 2010. Airway basal stem cells: a perspective on their roles in epithelial homeostasis and remodeling. Dis. Model. Mech. 3, 545—556.

Rompolas, P., Patel-King, R.S., King, S.M., 2009. Schmidtea mediterranea: a model system for analysis of motile cilia. Methods Cell Biol. 93, 81—98.

Sang, L., Miller, J.J., Corbit, K.C., Giles, R.H., Brauer, M.J., Otto, E.A., Baye, L.M., Wen, X., Scales, S.J., Kwong, M., et al., 2011. Mapping the NPHP-JBTS-MKS protein network reveals ciliopathy disease genes and pathways. Cell 145, 513—528.

Satir, P., 2016. Chirality of the cytoskeleton in the origins of cellular asymmetry. Philos. Trans. R. Soc. Lond. B Biol. Sci. 371, 20150408.

Sawamoto, K., Wichterle, H., Gonzalez-Perez, O., Cholfin, J.A., Yamada, M., Spassky, N., Murcia, N.S., Garcia-Verdugo, J.M., Marin, O., Rubenstein, J.L.R., et al., 2006. New neurons follow the flow of cerebrospinal fluid in the adult brain. Science 311, 629—632.

Schweickert, A., Walentek, P., Thumberger, T., Danilchik, M., 2011. Linking early determinants and cilia-driven leftward flow in left — right axis specification of *Xenopus laevis*: a theoretical approach. Differentiation 83, S67—S77.

Sedzinski, J., Hannezo, E., Wallingford, J.B., Tu, F., 2016. Emergence of an apical epithelial cell surface in vivo. 24—35.

Semenov, M.V., Habas, R., MacDonald, B.T., He, X., 2007. SnapShot: noncanonical Wnt signaling pathways. Cell 131.

Shimada, Y., Yonemura, S., Ohkura, H., Strutt, D., Uemura, T., 2006. Polarized transport of Frizzled along the planar microtubule arrays in *Drosophila* wing epithelium. Dev. Cell 10, 209—222.

Smith, J.C., Conlon, F.L., Saka, Y., Tada, M., 2000. Xwnt11 and the regulation of gastrulation in *Xenopus*. Philos. Trans. R. Soc. Lond. B Biol. Sci. 355, 923—930.

Song, R., Walentek, P., Sponer, N., Klimke, A., Lee, J.S., Dixon, G., Harland, R., Wan, Y., Lishko, P., Lize, M., et al., 2014. miR-34/449 miRNAs are required for motile ciliogenesis by repressing cp110. Nature 510, 115—120.

Sorokin, S.P., 1968. Reconstructions of centriole formation and ciliogenesis in mammalian lungs. J. Cell Sci. 3, 207—230.

Spassky, N., 2005. Adult ependymal cells are postmitotic and are derived from radial glial cells during embryogenesis. J. Neurosci. 25, 10—18.

Spemann, H., Mangold, H., 1924. In: Wilhelm Roux (Ed.), Ueber Induktion von Embryonalanlagen durch Implantation artfremder Organisatoren. Mikroskopische Anat. und Entwicklungsmechanik, 100.

St Johnston, D., Ahringer, J., 2010. Cell polarity in eggs and epithelia: parallels and diversity. Cell 141, 757—774.

Stubbs, J.L., Davidson, L., Keller, R., Kintner, C., 2006. Radial intercalation of ciliated cells during *Xenopus* skin development. Development 133, 2507—2515.

Tabler, J.M., Yamanaka, H., Green, J.B.A., 2010. PAR-1 promotes primary neurogenesis and asymmetric cell divisions via control of spindle orientation. Development 137, 2501—2505.

Tada, M., Smith, J.C., 2000. Xwnt11 is a target of *Xenopus* Brachyury: regulation of gastrulation movements via dishevelled, but not through the canonical Wnt pathway. Development 127, 2227—2238.

Tada, M., Concha, M.L., Heisenberg, C., 2002. Non-canonical Wnt signalling and regulation of gastrulation movements. Online 13, 251—260.

Tan, F.E., Vladar, E.K., Ma, L., Fuentealba, L.C., Hoh, R., Espinoza, F.H., Axelrod, J.D., Alvarez-buylla, A., Stearns, T., Kintner, C., et al., 2013. Myb promotes centriole amplification and later steps of the multiciliogenesis program. Development 140, 4277—4286.

Taschner, M., Bhogaraju, S., Lorentzen, E., 2011. Architecture and function of IFT complex proteins in ciliogenesis. Differentiation 1—11.

Tsang, W.Y., Dynlacht, B.D., 2013. CP110 and its network of partners coordinately regulate cilia assembly. Cilia 2, 9.

Turk, E., Wills, A.A., Kwon, T., Sedzinski, J., Wallingford, J.B., Stearns, T., 2015. Zeta-tubulin is a member of a conserved tubulin module and is a component of the centriolar basal foot in multiciliated cells. Curr. Biol. 25, 2177—2183.

Van Itallie, C.M., Anderson, J.M., 2014. Architecture of tight junctions and principles of molecular composition. Semin. Cell Dev. Biol. 36, 157—165.

Vladar, E.K., Brody, S.L., 2013. Analysis of Ciliogenesis in Primary Culture Mouse Tracheal Epithelial Cells, first ed. Elsevier Inc.

Vladar, E.K., Antic, D., Axelrod, J.D., 2009. Planar cell polarity signaling: the developing cell's compass. Cold Spring Harb. Perspect. Biol. 1, a002964.

Vladar, E.K., Bayly, R.D., Sangoram, A.M., Scott, M.P., Axelrod, J.D., 2012. Microtubules enable the planar cell polarity of airway cilia. Curr. Biol. 22, 2203—2212.

Vladar, E.K., Lee, Y.L., Stearns, T., Axelrod, J.D., 2015. Observing planar cell polarity in multiciliated mouse airway epithelial cells. Methods Cell Biol. 127, 37—54.

Vladar, E.K., Nayak, J.V., Milla, C.E., Axelrod, J.D., 2016. Airway epithelial homeostasis and planar cell polarity signaling depend on multiciliated cell differentiation. JCI Insight 1, 1—18.

Walentek, P., Quigley, I.K., 2017. What we can learn from a tadpole about ciliopathies and airway diseases : using systems biology in *Xenopus* to study cilia and mucociliary epithelia. Genes. J. Genet. Dev. 1—13.

Walentek, P., Schneider, I., Schweickert, A., Blum, M., 2013. Wnt11b is involved in cilia-mediated symmetry breakage during *Xenopus* left-right development. PLoS One 8, e73646.

Walentek, P., Bogusch, S., Thumberger, T., Vick, P., Dubaissi, E., Beyer, T., Blum, M., Schweickert, A., 2014. A novel serotonin-secreting cell type regulates ciliary motility in the mucociliary epidermis of *Xenopus* tadpoles. Development 141, 1526—1533.

Walentek, P., Beyer, T., Hagenlocher, C., Müller, C., Feistel, K., Schweickert, A., Harland, R.M., Blum, M., 2015. ATP4a is required for development and function of the *Xenopus* mucociliary epidermis — a potential model to study proton pump inhibitor-associated pneumonia. Dev. Biol. 408, 292—304.

Walentek, P., Quigley, I.K., Sun, D.I., Sajjan, U.K., Kintner, C., Harland, R.M., 2016. Ciliary transcription factors and miRNAs precisely regulate Cp110 levels required for ciliary adhesions and ciliogenesis. Elife 5, 1—24.

Wallingford, J.B., Mitchell, B., 2011. Strange as it may seem: the many links between Wnt signaling, planar cell polarity, and cilia. Genes Dev. 25, 201—213.

Wallingford, J.B., 2006. Planar cell polarity, ciliogenesis and neural tube defects. Hum. Mol. Genet. 15, 227—234.

Wallingford, J.B., 2010. Planar cell polarity signaling, cilia and polarized ciliary beating. Curr. Opin. Cell Biol. 22, 597—604.

Wallmeier, J., Al-Mutairi, D.A., Chen, C.-T., Loges, N.T., Pennekamp, P., Menchen, T., Ma, L., Shamseldin, H.E., Olbrich, H., Dougherty, G.W., et al., 2014. Mutations in CCNO result in congenital mucociliary clearance disorder with reduced generation of multiple motile cilia. Nat. Genet. 46, 646—651.

Wang, Y., Nathans, J., 2007. Tissue/planar cell polarity in vertebrates: new insights and new questions. Development 134, 647—658.

Werner, M.E., Mitchell, B.J., 2012. Understanding ciliated epithelia: the power of *Xenopus*. Genesis 50, 176—185.

Werner, M.E., Hwang, P., Huisman, F., Taborek, P., Yu, C.C., Mitchell, B.J., 2011. Actin and microtubules drive differential aspects of planar cell polarity in multiciliated cells. J. Cell Biol. 195, 19—26.

Werner, M.E., Mitchell, J.W., Putzbach, W., Bacon, E., Kim, S.K., Mitchell, B.J., 2014. Radial intercalation is regulated by the par complex and the microtubule-stabilizing protein CLAMP/Spef1. J. Cell Biol. 206, 367—376.

Wu, J., Roman, A.-C., Carvajal-Gonzalez, J.M., Mlodzik, M., 2013. Wg and Wnt4 provide long-range directional input to planar cell polarity orientation in *Drosophila*. Nat. Cell Biol. 15, 1045—1055.

Yang, Y., Mlodzik, M., 2015. Wnt-frizzled/planar cell polarity signaling: cellular orientation by facing the wind (Wnt). Annu. Rev. Cell Dev. Biol. 31, 623—646.

Yasunaga, T., Hoff, S., Schell, C., Helmstädter, M., Kretz, O., Kuechlin, S., Yakulov, T.A., Engel, C., Müller, B., Bensch, R., et al., 2015. The polarity protein inturned links NPHP4 to daam1 to control the subapical actin network in multiciliated cells. J. Cell Biol. 211, 963—973.

Yee, L.E., Reiter, J.F., 2015. Ciliary vesicle formation: a prelude to ciliogenesis. Dev. Cell 32, 665—666.

Zariwala, M.A., Omran, H., Ferkol, T.W., 2011. The emerging genetics of primary ciliary dyskinesia. Proc. Am. Thorac. Soc. 8, 430—433.

Zhang, S., Mitchell, B.J., 2015. Centriole biogenesis and function in multiciliated cells. In: Methods in Cell Biology. Elsevier, pp. 103—127.

Zhang, S., Mitchell, B.J., 2016. Basal bodies in *Xenopus*. Cilia 5, 1—6.

Zhao, H., Zhu, L., Zhu, Y., Cao, J., Li, S., Huang, Q., Xu, T., Huang, X., Yan, X., Zhu, X., 2013. The Cep63 paralogue Deup1 enables massive de novo centriole biogenesis for vertebrate multiciliogenesis. Nat. Cell Biol. 15, 1434—1444.

Index

Note: 'Page numbers followed by "f" indicate figures, "t" indicate tables and "b" indicate boxes.'

O

P

Printed and bound by CPI Group (UK) Ltd, Croydon, CR0 4YY

08/05/2025

01864999-0001